IRRIGATION AND DRAINAGE PERFORMANCE ASSESSMENT

Practical Guidelines

 This book is the result of long-term cooperation between the members of the ICID Working Group on Performance Assessment of Irrigation and Drainage.

The members of the Working Group were:
Dr M.G. Bos, Chairman (The Netherlands)
Dr Fatma Abdel-Rahman Attia (Egypt)
Dr M.N. Bhutta (Pakistan)
Professor N. Borin (Italy)
Dr R.A.L. Brito (Brazil)
Dr M.A. Burton (UK)
Mr J. Chambouleyron (Argentina)
Mr Lee Chang Chi (Taiwan RoC)
M.J.F. Metzger (Canada)
Dr D.J. Molden (USA)
Dr B. Molle (France)
Mr J.A. Ortiz F.-Urrutia (Spain)
Mr J. Plantey (France)
Dr Sang Hyun Park (South Korea)
Dr T. Watanabe (Japan)
Mr G.J. Wright (Australia)
Dr J. Yang (P.R. China)
Mr I. Makin for IWMI (Sri Lanka)

The authors can be reached at the following institutions:

 International Institute for Land Reclamation and Improvement Alterra-ILRI,
PO Box 47, 6700 AA Wageningen,
The Netherlands
www.ilri.nl

 ITAD~Water Ltd, 12 English Business Park,
English Close, Hove,
West Sussex BN3 7ET, UK
www.itad.com

 International Water Management Institute,
PO Box 2075, Colombo, Sri Lanka
www.iwmi.cgiar.org

IRRIGATION AND DRAINAGE PERFORMANCE ASSESSMENT
Practical Guidelines

M.G. Bos

*International Institute for Land Reclamation and Improvement
Wageningen, The Netherlands*

M.A. Burton

*ITAD~Water
Hove, UK*

and

D.J. Molden

*International Water Management Institute
Colombo, Sri Lanka*

CABI Publishing

CABI Publishing is a division of CAB International

CABI Publishing
CAB International
Wallingford
Oxfordshire OX10 8DE
UK

Tel: +44 (0)1491 832111
Fax: +44 (0)1491 833508
E-mail: cabi@cabi.org
Website: www.cabi-publishing.org

CABI Publishing
875 Massachusetts Avenue
7th Floor
Cambridge, MA 02139
USA

Tel: +1 617 395 4056
Fax: +1 617 354 6875
E-mail: cabi-nao@cabi.org

© M.G. Bos, M.A. Burton and D.J. Molden 2005. All rights reserved. No part of this publication may be reproduced in any form or by any means, electronically, mechanically, by photocopying, recording or otherwise, without the prior permission of the copyright owners.

A catalogue record for this book is available from the British Library, London, UK.

Library of Congress Cataloging-in-Publication Data

Bos, Marinus Gijsberthus, 1943–
   Irrigation and drainage performance assessment : practical guidelines / M.G. Bos, M.A. Burton, and D.J. Molden.
      p. cm.
   Includes bibliographical references and index.
   ISBN 0-85199-967-0 (alk. paper)
   1. Irrigation--Evaluation. 2. Irrigation--Management. 3. Drainage--Evaluation.
I. Burton, M. A. S. II. Molden, D. J. III. Title.

TC805.B683 2004
631.5′87--dc22

2004013236

ISBN 0 85199 967 0

Typeset by Columns Design Ltd, Reading
Printed and bound in the UK by Cromwell Press, Trowbridge

Contents

Abstract		vi
Preface		vii
Chapter 1	Introduction	1
Chapter 2	Framework for Performance Assessment	6
Chapter 3	Performance Indicators for Irrigation and Drainage	26
Chapter 4	Operational and Strategic Performance Assessment	62
Chapter 5	Diagnosing Irrigation Performance	87
Chapter 6	Data Management for Performance Assessment	117
Appendix 1		140
Appendix 2		143
Index		155

Abstract

Performance assessment of irrigation and drainage is the systematic observation, documentation and interpretation of the management of an irrigation and drainage system, with the objective of ensuring that the input of resources, operational schedules, intended outputs and required actions proceed as planned. Following an introduction on this concept in Chapter 1, a framework on performance assessment is presented in Chapter 2. Chapter 3 then defines and discusses 23 recommended indicators covering the water balance, the environment and the economic aspects of the region. The method by which the indicators are used for operational and strategic performance assessment is discussed in Chapter 4, while the diagnostic use of the concept is presented in Chapter 5. The ultimate purpose of performance assessment is to achieve an efficient and effective project performance by providing relevant feedback to the project management at all levels. All related aspects of data management and communication with water users are illustrated in Chapter 6.

Preface

The purpose of this handbook is to draw together the knowledge that has been gained in irrigation and drainage performance assessment over the last 10–15 years. Much has been written, and it is time to put down guidelines to enable practitioners to apply the processes and procedures that have been developed. The handbook aims to provide a generic framework for performance assessment with guidance on how to design and carry out performance assessment programmes.

The handbook is aimed at a variety of irrigation and drainage professionals, including scheme managers, researchers and consultants. Performance assessment is an essential management task. If the use of water for irrigation is to be improved, then we must understand current levels of performance and identify measures for improvement.

The primary advantage of 'performance-oriented water management' is that a water management strategy has to be formulated in consultation with all stakeholders. This 'agreed strategy' then forms the foundation of the operational rules of the irrigation and drainage project. In addition, performance assessment has the following major advantages:

- The use and consumption of water by different user groups (farming, drinking water, industrial water, ecosystems, etc.) can be quantified and weighted with respect to each other and the actual water use policy.
- Time series of performance indicators that are plotted with respect to the related critical value or benchmark value show the time when these values will be reached. This then defines the period that is available for management actions.
- The use of various resources (land, water, funds, knowledge, etc.) for the production of food and fibre are quantified and compared with target values. Negative effects can be diagnosed.

- The impact of management actions on the use of resources and on crop yield can be monitored with respect to target values.
- The systematic presentation of the actual performance of the irrigation and/or drainage project improves communication between stakeholders.

Members of the ICID (International Commission on Irrigation and Drainage) Working Group on Performance Assessment of Irrigation and Drainage have field-tested these guidelines through several case studies. These studies showed that the range of potential applications for these guidelines is unlimited. Acknowledgements are due to all members of the Working Group for their contributions and comments, in particular Jacques Plantey, Bruno Molle and Ricardo Brito, and the USCID Working Group on Irrigation Performance Assessment.

We hope that this book will contribute to the effective management of one of the earth's most widely needed, used and wasted natural resources: water.

Marinus G. Bos
Wageningen, The Netherlands

Martin A. Burton
Hove, UK

David J. Molden
Colombo, Sri Lanka

1 Introduction

Background

Performance of irrigated agriculture must improve to provide additional food to a growing and more affluent population, but it is constrained by water scarcity and the resulting competition for scarce water resources. More food will have to come from existing large-scale irrigation works, from drained lands, from privately owned groundwater wells and small-scale systems delivering supplemental irrigation, and often with less water.

Withdrawals for irrigation have increased dramatically, especially during 1950–1980, when many large infrastructure projects were constructed. There is no doubt that irrigation development contributed to national food security, to economic development and to the relief of poverty. Yet, many problems remain – natural ecosystems have borne the burden of this development of water for agriculture, and malnutrition remains, with 790 million people estimated to live with hunger.

Concerns and problems over water scarcity will certainly affect irrigated agriculture. The rate of increase in irrigation withdrawals will not be the same as over the last 25-year period. From 1995 to 2025, FAO forecasts a growth in irrigation withdrawals of 14%, while IWMI sees a 17% growth in withdrawals for irrigation. But food production from irrigated lands during the same period should grow by at least 40% to meet the needs of a 33% increase in population, and to satisfy trends for improved nutrition.

There is increasing competition for water. Water is increasingly being transferred from irrigated agriculture to higher valued industrial and urban uses, and irrigated land is going out of production from urban sprawl. Water quality problems increase with rising industrialization and inefficient irrigation water use, leading to pollution and salinization. There is a call for more water to be reserved for environmental uses. It is not clear how much land is going out of production due to salinization, but it is clearly a threat to irrigated food production systems.

The challenges are set out before us. Society is demanding much more from irrigated agriculture. First, we have to produce more food per unit of water available for agriculture. Second, we have to do this without further environmental degradation. Third, we have to target the needs of increased income for farmers, and to reduce levels of poverty, especially in the developing world. It is a task for water managers, farmers and all involved in the irrigation and drainage sector.

A starting point is improving how water is managed within irrigation and drainage systems. We know that many irrigation systems are performing below their capability. An adequate service is not provided to farmers, resulting in inequitable and unreliable distribution. Water productivity is below acceptable levels, and irrigation activities contribute to pollution. Of course, there are many systems that perform well, where lessons are to be learned, but there are also many poorly performing irrigation systems, where performance improvements can be made.

Overview

Performance assessment in irrigation and drainage can be defined as the systematic observation, documentation and interpretation of activities related to irrigated agriculture with the objective of continuous improvement. Performance assessment is an activity that supports the planning and implementation process. The ultimate purpose of performance assessment is to achieve an efficient and effective use of resources by providing relevant feedback to the scheme management at all levels. As such, it may assist the scheme management in determining whether the performance is satisfactory and, if not, which and where corrective or different actions need to be taken in order to remedy the situation. It should provide insights into the process of irrigation and drainage so that managers, farmers and planners can do business in new, more productive and efficient ways.

A systematic and timely flow of actual (measured or collected) data on key aspects of a scheme is an essential condition for the monitoring of performance to become an effective management tool. These data should provide sufficient information for the managers to answer two simple questions (Murray-Rust and Snellen, 1993):

- 'Am I doing things right?', a question that asks whether the intended level of service or operation that has been set (or agreed upon) is being achieved. This is the basis for good *operational* performance.
- 'Am I doing the right thing?', a question that aims at finding out whether the wider objectives of irrigation and drainage are being fulfilled, and fulfilled efficiently. The latter is part of the process of assessment of *strategic* performance.

Operational performance is concerned with the routine implementation of operational procedures based on fixed or negotiated service specifications. It specifically measures the extent to which intentions or target levels are being met at any moment in time, at every considered level of the scheme and thus requires the actual inputs of resources and the related outputs to be measured.

Strategic performance is a longer-term activity that assesses the extent to which all available resources have been utilized to achieve the service or operational level efficiently, and explores whether achieving this service or operation also meets the broader set of objectives. A time-series of the indicator and its rate of change are commonly used in this activity. Strategic performance is used to revise longer-term goals, overall operational procedures to meet the changing demands of farmers, managers and society.

Available resources in this context refer not merely to financial resources; they also cover the natural resource base (land and water) and the human resources provided to operate, maintain and manage irrigation and drainage systems. Strategic management involves not only the system manager, but also higher level staff in agencies at the national planning and policy levels.

Application of Performance Assessment

Performance assessment can be used in a variety of ways, including:

- Operational performance assessment by scheme managers to determine how the operational processes are performing. The processes studied could relate to the overall production, or they could be broken down into sub-processes such as main system water delivery, on-farm water delivery, crop production, etc. depending on the level at which the analysis is required.
- Strategic performance assessment by government or scheme owners to understand how a scheme or schemes are performing and using available resources.
- Diagnostic performance assessment for scheme managers to understand the causes of low or high performance.
- Diagnostic performance assessment for consultants as a prelude to design and implementation of interventions for system improvement and rehabilitation.
- Performance assessment on behalf of government or other agencies to monitor how systems are satisfying identified objectives.
- Performance assessment and diagnostic analysis by research organizations to understand generic causes of low or high performance on irrigation schemes.
- Comparative performance assessment to compare performance of one scheme with another in order to set appropriate benchmark standards or identify processes that lead to higher performance.

- Benchmarking of scheme(s) with other scheme(s) exhibiting good performance, thereby identifying best practice.

Performance assessment is possible on any type of irrigation and drainage scheme, ranging from large scale to small scale, commercial estate to farmer-managed, highly technical systems with computer control of gates to simple gravity-fed proportional division systems. The detail of the performance assessment changes for each case, the overall process does not.

The application of performance assessment procedures will vary depending on the purpose of the assessment and the type of scheme. The assessment required for a large centrally managed commercial estate type scheme will differ from that for a smallholder farmer-managed irrigation scheme, the assessment possible on a system equipped with measuring structures at division points will differ from that in a system where there are no measuring structures.

It is intended that these guidelines will help in the planning and implementation of performance assessment programmes in the wide variety of situations that exist in irrigated agriculture worldwide.

Structure of the Guidelines

In Chapter 2, a generic framework is outlined for planning and implementing performance assessment programmes, be they for short-term one-off research studies or long-term programmes used by scheme managers to monitor and evaluate seasonal and annual performance of an irrigation and drainage scheme over many years. This chapter introduces the theory and process of performance assessment, and serves to provide a context for subsequent chapters. Chapter 3 builds on Chapter 2 by providing detailed information about the indicators that can be used to assess performance within a scheme. It looks at the characteristics, types and nature of performance indicators before providing guidance on the selection of performance indicators for a particular situation. While it is not possible to provide specific guidance on performance indicators to be used in individual cases, a selected number of commonly used indicators are identified and defined. This selected set is supported by an extended set in Appendix 2, together with references to their origin and application. Chapter 4 describes the use of performance assessment by scheme management with a view to improving the level of service provision and overall performance. The chapter outlines service-oriented management for different types of schemes and shows how performance assessment can be applied in a practical context, within the constraints of available time, money and personnel. Chapter 5 follows on with guidelines for diagnostic analysis of irrigation and drainage schemes, outlining procedures that can be used to identify and relieve performance constraints within a scheme. Chapter 6 deals with the data management required for performance assessment, with emphasis on the accuracy and cost-effectiveness of the data collected.

Reference

Murray-Rust, D.H. and Snellen, W.B. (1993) *Irrigation System Performance Assessment and Diagnosis.* Joint IIMI/ILRI/IHEE Publication. International Irrigation Management Institute, Columbo, Sri Lanka.

2 Framework for Performance Assessment

Introduction

This chapter sets out a generic framework for performance assessment within the irrigation and drainage sector. Subsequent chapters build on this framework, adding detail to the various stages involved in implementing a performance assessment programme.

All performance assessment programmes require a framework to define and guide the work. Several frameworks have been proposed in the past. In some cases these have been specific to a particular scheme, in other cases they have been more generic. Key players in the formulation of generic frameworks have been Bottrall (1981), Abernethy (1984), Oad and McCornick (1989), Svendsen (1990), Small and Svendsen (1992) and Murray-Rust and Snellen (1993). The framework described herein builds on this previous work, and work by Burton and Mututwa (2002) and Mututwa (2002).

The framework serves to define why the performance assessment is needed, what data are required, what methods of analysis will be used, who will use the information provided, etc. Without a suitable framework the performance assessment programme may fail to collect all the necessary data, and may not provide the required information and understanding.

The framework is based on a series of questions (Fig. 2.1). The first stage, 'purpose and strategy', looks at the broad scope of the performance assessment – who it is for, from whose viewpoint it is undertaken, who will carry it out, its type and extent. Once these issues are decided, the performance assessment programme can be designed, selecting suitable criteria for the performance assessment, performance indicators and the data that will be collected. The implementation of the planned programme follows, with data being collected, processed and analysed. The final part of the programme is to act on the information provided, with a variety of actions possible, ranging from changes to long-term goals and strategy, to improvements in day-to-day procedures for system management, operation and maintenance.

The framework can also be applied to monitor and evaluate existing performance assessment programmes, in a similar manner to that in which the logical framework[1] is used to monitor and evaluate project performance.

Purpose and Strategy

The initial part of formulating a performance assessment programme is to decide on the purpose and scope of the performance assessment. Key issues relate to who the assessment is for, from whose viewpoint, the type of assessment and the extent/boundaries. It is important that adequate time is spent on this part of the work as it structures the remaining stages.

Purpose

As with any project or task, it is essential that the purpose and objectives of the performance assessment be defined at the outset.

Three levels of objective-setting can be identified:

1. Rationale.
2. Overall objective.
3. Specific objectives.

The *rationale* outlines the reason for which a performance assessment programme is required. The *overall objective* details the overall aim of the performance assessment programme, while the *specific objectives* provide further detail on how the overall objective will be achieved (Table 2.1).

Establishing the rationale and identifying the overall and specific objectives of the performance assessment programme is not always straightforward; care needs to be taken at this stage of planning to ensure that these objectives are clearly defined before proceeding further.

For whom?

The performance assessment can be carried out on behalf of a variety of stakeholders. These include:

- Government.
- Funding agencies.

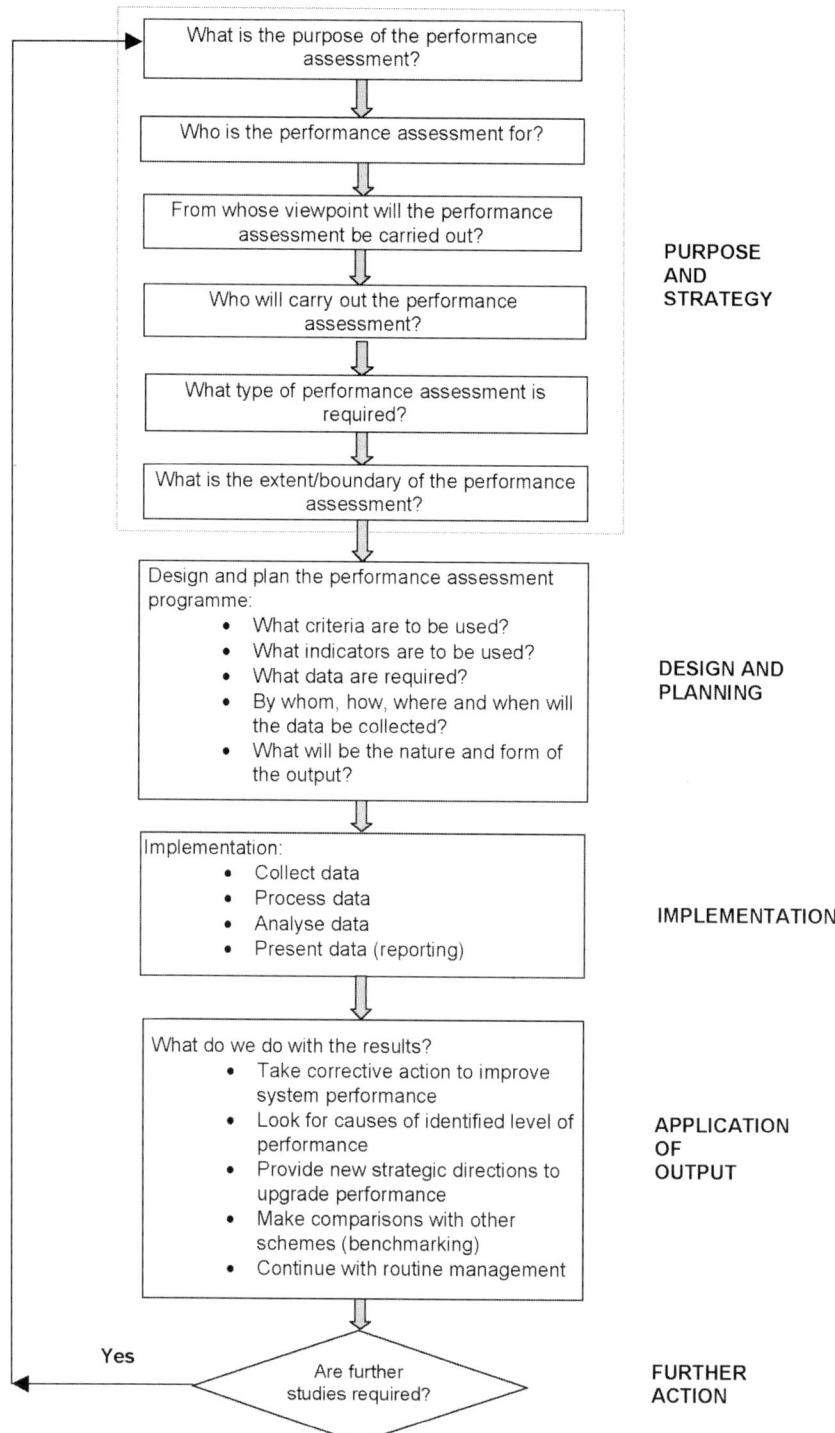

Fig. 2.1. Framework for performance assessment of irrigation and drainage schemes.

Table 2.1. Example of the rationale and a set of objectives for a performance assessment programme.

Rationale:	Water management needs to be improved if all farmers within the scheme[a] are to obtain adequate livelihoods
Overall objective:	To identify feasible and sustainable water management practices which lead to improved crop production and thereby income for the farming community
Specific objectives:	Monitor water demands and allocations at all control points (primary, secondary and tertiary canal intakes)
	Analyse current match between water supply and demand, and identify areas for improvement
	Formulate strategy for improvement
	Implement strategy
	Monitor and evaluate impact

[a]The term 'irrigation and drainage system' refers to the network of irrigation and drainage channels, including structures. The term 'irrigation and drainage scheme' refers to the total irrigation and drainage complex, the irrigation and drainage (I&D) system, the irrigated land, villages, roads, etc.

- Irrigation and drainage service providers.
- Irrigation and drainage system managers.
- Farmers.
- Research organizations.

Who the assessment is for is closely linked to the purpose of the assessment.

From whose viewpoint?

The assessment may be carried out on behalf of one stakeholder or group of stakeholders, but may be looking at performance assessment from the perspective of another stakeholder or group of stakeholders. Government may commission a performance assessment, for example, to be carried out by a research institute to study the impact of system performance on farmer livelihoods. Farmers might commission a study of the irrigation service provider in order to ascertain if they are receiving an adequate level of service for fees paid.

By whom?

Different organizations or individuals have different capabilities in respect of performance assessment, and different types of performance assessment will require different types of organization or individuals to carry out the assessment (Table 2.2). A scheme manager might establish a

Table 2.2. Examples of for whom, from whose viewpoint and by whom performance assessment might be carried out.

For whom?	From whose viewpoint?	By whom?
Scheme manager	The scheme management	Scheme manager and staff
Government	Government (for example: return on proposed investment)	Consultant
Government	Society in general, but specifically water users	Government regulatory authority
Funding agency	Farmers (livelihood)	Consultant
Scientific community	The management of the system	Research institute/university
Farmers	Farmers	Consultant

performance assessment programme using existing operation and maintenance (O&M) personnel to be able to monitor and evaluate scheme performance. A government agency might employ a consultant to carry out performance assessment of a scheme with a view to further investment, while a university research team might carry out a research programme to identify and understand generic factors that affect system performance.

Type

Small and Svendsen (1992) identify four different types of performance assessment, to which a fifth, diagnostic analysis, can be added:

1. Operational.
2. Accountability.
3. Intervention.
4. Sustainability.
5. Diagnostic analysis.

The type of performance assessment is linked with the purpose; in fact Small and Svendsen refer to these categories as the rationale for performance assessment.

Operational performance assessment relates to the day-to-day, season-to-season monitoring and evaluation of scheme performance. *Accountability* performance assessment is carried out to assess the performance of those responsible for managing a scheme. *Intervention* assessment is carried out to study the performance of the scheme and, generally, to look for ways to enhance that performance. Performance assessment associated with *sustainability* looks at the longer-term resource use and impacts. *Diagnostic analysis* seeks to use performance assessment to track down the cause, or causes, of performance in order that improvements can be made or performance levels sustained. Chapter 5 covers diagnostic and intervention assessment in more detail.

Internal or external assessment

It is important to define at the outset whether the performance assessment relates to one scheme (internal analysis) or comparison between schemes (external analysis).

A significant problem with performance assessment of irrigation and drainage schemes is the complexity and thus variety of types of scheme. This makes comparison between schemes problematic. Some schemes are farmer-managed, some are private estates with shareholders, some are gravity-fed, some fed via pressurized pipe systems, etc. There is as yet no definitive methodology for categorizing irrigation and drainage schemes, therefore there will always be discussion as to whether one is comparing like with like. A shortlist of key descriptors for irrigation and drainage schemes is presented in Table 2.3, with an extended list being provided in Appendix 1. This list of descriptors can be used to select schemes with similar key characteristics for comparison.

It is important to understand, however, that comparison between different types of scheme can be equally valuable, as for instance might be the case for governments in comparing the performance of privately owned estates with smallholder irrigation schemes. The two have different management objectives and processes, but their performance relative to criteria based on the efficiency and productivity of resource use (land, water, finance, labour) would be of value in policy formulation and financial resource allocation.

Benchmarking of irrigation and drainage systems is a form of comparative (external) performance assessment that is increasingly being used. Benchmarking seeks to compare the performance of 'best practice' systems with the currently assessed system, and to understand where the differences in performance lie. Initially performance assessment might be focused on a comparison of output performance indicators (water delivery, crop production, etc.), followed by diagnostic analysis to understand: (i) what causes the relative difference in performance, and (ii) what measures can feasibly be taken to raise performance in the less well-performing system(s).

The selection of performance assessment criteria will be influenced by whether the exercise looks internally at the specific objectives of an irrigation scheme, or whether it looks to externally defined performance criteria. Different schemes will have different objectives, and different degrees to which these objectives are implicitly or explicitly stated. It may well be that when measured against its own explicitly stated objectives (for example, to provide 1000 people with secure livelihoods) a scheme is deemed a success. However, when measured against an external criterion of crop productivity per unit of water used, or impact on the environment, it may not perform as well. This reinforces the point made earlier that assessment of performance is often dependent on people's perspective – irrigation is seen as beneficial by farmers, possibly less so by fishermen and downstream water users.

Table 2.3. Key descriptors for irrigation and drainage schemes.

Descriptor	Possible options	Explanatory notes	Example
Irrigable area	–	Defines whether the scheme is large, medium or small scale	8567 ha
Annual irrigated area	Area supplied from surface water; Area supplied from groundwater	Shows the intensity of land use and balance between surface or groundwater irrigation	7267 ha; 4253 ha surface; 3014 ha groundwater
Climate	Arid; semi-arid; humid tropics; Mediterranean	Sets the climatic context. Influences the types of crops that can be grown	Mediterranean
Average annual rainfall (P)	–	Associated with climate, sets the climatic context and need for irrigation and/or drainage	440 mm
Average annual reference crop evapotranspiration (ET_o)	–	Associated with climate, sets the climatic context and need for irrigation	780 mm
Water source	Storage on river; groundwater; run-of-the river; conjunctive use of surface and groundwater	Describes the availability and reliability of irrigation water supply	Over-year storage reservoir in upper reaches; Groundwater aquifers
Method of water abstraction	Pumped; gravity; artesian	Influences the pattern of supply and cost of irrigation water	Gravity-fed from rivers, pumped from groundwater
Water delivery infrastructure	Open channel; pipelines; lined; unlined	Influences the potential level of performance	Open channel, lined primary and secondary canals
Type of water distribution	Demand; arranged on demand; arranged; supply oriented	Influences the potential level of performance	Arranged on demand
Predominant on-farm irrigation practice	Surface: furrow, level basin, border, flood, ridge-in-basin; Overhead: rain-gun, lateral move, centre pivot; drip/trickle; Subsurface: drip	Influences the potential level of performance	Predominantly furrow, with some sprinkler and (increasingly) drip

Table 2.3. *Continued.*

Descriptor	Possible options	Explanatory notes	Example
Major crops (with percentages of total irrigated area)	–	Sets the agricultural context. Separates out rice and non-rice schemes, monoculture from mixed cropping schemes	Cotton (53%) Grapes (27%) Maize (17%) Other crops (3%)
Average farm size	–	Important for comparison between schemes, whether they are large estates or smallholder schemes	0.5–5 ha (20%) >5–20 ha (40%) >20–50 ha (20%) >50 ha (20%)
Type of management	Government agency; private company; joint government agency/farmer; farmer-managed	Influences the potential level of performance	River system – government; primary and secondary systems – water users' associations

Extent/boundaries

The extent of the performance assessment needs to be identified and the boundaries defined. Two primary boundaries relate to spatial and temporal dimensions. *Spatial* relates to the area or number of schemes covered (is the performance assessment limited to one secondary canal within a system, to one system, or to several systems); *temporal* relates to the duration of the assessment exercise and temporal extent (1 week, one season, or several years).

Other boundaries are sometimes less clear cut, and can relate to whether the performance assessment aims to cover technical aspects alone, or whether it should include institutional and financial aspects. How much influence, for example, does the existence of a water law on the establishment of water users' associations have on the performance of transferred irrigation and drainage systems?

The use of the *systems approach* advocated by Small and Svendsen (1992) can add to the definition and understanding of the boundaries and extent of the performance assessment programme. The systems approach focuses on *inputs, processes, outputs* and *impacts*. Measurement of outputs (for example, water delivery to tertiary unit intakes) provides information on the effectiveness of the use of inputs (water abstracted at river intake), while comparison of outputs to inputs provides information on the efficiency of the process of converting inputs into outputs. The process of transforming inputs into outputs has impacts down the line – the pattern of water delivery to the tertiary intake has, for example, an impact on the level of crop production attained by the farmer.

Measurements of canal discharges will provide information on how the irrigation system (network) is performing, but tell us little about the performance of the irrigation and drainage scheme as a whole. To obtain this information we need to collect data within the irrigated agriculture system, and the agricultural economic system to set the performance of the irrigation system in context. Care is needed here in relating the performance of the irrigation system (e.g. adequate and timely water supply) to that of the agricultural economic system (e.g. farmer income) as many variables intervene between the supply of the irrigation water and the money received by the farmer for the crops produced.

Alternative systems can be drawn up, as, for example, that shown in Fig. 2.2 linking the performance of irrigation and drainage into the wider institutional context.

Governing the selection of the criteria and performance indicators used in the exercise will be decisions on the 'systems' which need to be included in the performance assessment exercise, and the related components (inputs, processes, outputs or impacts). The performance assessment programme may be interested in the level of outputs (crop production), and also the efficiency of resource use (production per unit of land, water, finance, labour, etc.). It might also be interested in the processes (e.g. canal conveyance efficiency). Impacts might relate to complying with statutory regulations or protection of the environment (e.g. salinity levels of drainage water). It is not necessary that all systems or system stages are studied, it is important, however, to be aware of the context in which a given performance assessment programme is set.

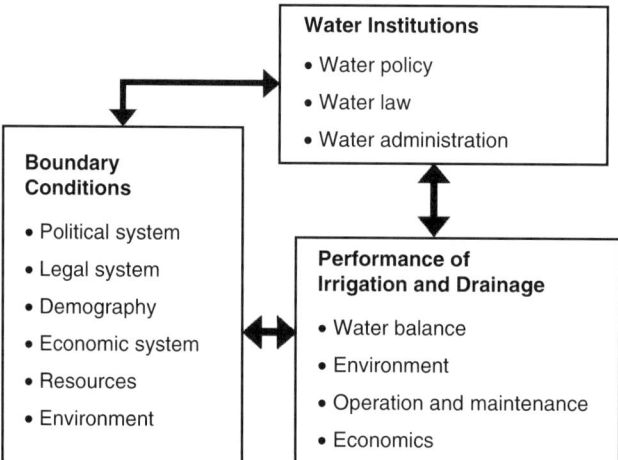

Fig. 2.2. Irrigation and drainage performance in relation to the wider institutional context.

Design of the Performance Assessment Programme

Having specified the approach to the performance assessment programme in terms of the purpose and strategy, the performance assessment programme can be designed. The key issues to consider are:

- What criteria are to be used?
- What performance indicators are to be used?
- What data are required?
- By whom, how, where and when will the data be collected?
- What is the required form of output?

Performance criteria and scheme objectives

In the literature the terms performance criteria, performance indicators and performance measures are used by different authors to mean different things. The following definitions are proposed in order to clarify the terms *performance criteria*, *objectives*, *performance indicators* and *targets*:

1. Objectives are made up of criteria: (i) 'To maximize agricultural *production*'; (ii) 'To ensure *equity* of water supply to all farmers'; and (iii) 'To optimize the *efficiency* of water distribution'.
2. Criteria can be measured using *performance indicators*.
3. Defined *performance indicators* identify data requirements.
4. Data can then be collected, processed and analysed.
5. If *target*, *standards*, *reference* or *benchmark* values of *performance indicators* are set or known then performance can be assessed.

In selection of criteria for performance assessment it is necessary to define whether the assessment will be made against the scheme's stated objectives and criteria, or against an alternative set of performance objectives or criteria. An example of where a scheme's objectives and target values are stated is shown in Table 2.4. In this case the targets for cropped area and crop production (in terms of crop production and value) can also be monitored over time to assess the sustainability of the scheme.

While an irrigation scheme may have stated objectives, its performance may need to be assessed against different criteria (Table 2.5). For example, a government might assess a scheme's performance in relation to the country's economic needs, or environmental sustainability and impact. Simply because these criteria are not stated in the objectives for the scheme does not mean that the scheme cannot be assessed against such externally stipulated criteria. For example, a scheme may have no stated objectives about pollution loading, but an environmental regulatory agency may have their own standards against which the scheme's performance is assessed.

In some of the literature on performance assessment, authors have stated that performance should be assessed against objectives set for a given scheme. This is an obvious starting point, but, as found by Ijir and

Table 2.4. Example of linkage of objectives, criteria, performance indicators and targets. Source: Calculations for Mogambo Irrigation Scheme, Somalia in Burton (1993).

Objective	Criterion	Performance indicator	Target value
Maximize area harvested	Productivity	Cropping intensity	2052 ha (100%)
Maximize total crop production	Productivity	Total production	7600 t
Maximize total value of agricultural production	Productivity	Total value of production	$1,067,238
Maximize productivity of water	Productivity	Water productivity Value of production per unit water	0.16 kg/m^3 $0.023/m^3
Maximize equity of water supply	Equity	Area planted/area harvested Delivery performance ratio	1.0 SD < 10%

Burton (1998), this approach fails when there are no explicitly stated objectives for the scheme.

As outlined in Murray-Rust and Snellen (1993), the setting of objectives is a crucial part of the management process, and much has been written on the subject in the context of business management. Some key points in relation to objective setting for irrigation management and performance assessment are outlined below:

1. *Explicit or implicit.* Objectives can be *explicit*, where they are clearly stated, or *implicit*, where they are assumed rather than stated. For example, for the Ganges Kobadak irrigation scheme in Bangladesh the explicit objective is food production, but an (essential) implicit objective is flood protection to prevent the irrigation scheme being inundated by the waters of the Ganges River. In performance assessment it is important to identify both types of objectives.

2. *Hierarchy of objectives.* Objectives occur at different levels within a system or systems. A hierarchy of objectives for irrigation development, identified by Sagardoy *et al.* (1982), was, in ascending order:

- Appropriate use of water.
- Appropriate use of agricultural inputs.
- Remunerative selling of agricultural products.
- Improvement in social facilities.
- Betterment of farmers' welfare.

Each of these objectives is important at its own system level, satisfying the objectives at one level means that those at another (higher) level might also be satisfied. This hierarchy of objectives is an integral part of the *Logical Framework* project planning tool, moving from outputs to purpose to satisfy the overall goal.

3. *Ranking or weighting of objectives.* Within a system there may be several, sometimes competing objectives. For performance assessment these

Table 2.5. Criteria for good system performance according to type of person (Chambers, 1988).

Type of person	Possible first criterion of good system performance
Landless labourer	Increased labour demand, days of working and wages
Farmer	Delivery of an adequate, convenient, predictable and timely water supply
Irrigation engineer	Efficient delivery of water from headworks to the tertiary outlet
Agricultural economist	High and stable farm production and incomes
Economist	High internal rate of return
Political economist	Equitable distribution of benefits, especially to disadvantaged groups

may need to be ranked or weighted and assessments made to evaluate how well individual and collective objectives are satisfied. This process is commonly termed multi-criteria analysis. An example of the weightings and rankings attached to individual objectives, depending on whether the irrigation scheme is run as a state farm or settlement scheme, are presented in Table 2.6. Objectives to maximize equitable distribution of water are favoured for a settlement scheme, while objectives to maximize value of production are favoured for a state farm. A similar approach to analysis incorporating competing performance criteria has been adopted by Molden and Gates (1990) and Burt and Styles (1999).

Performance indicators

Performance is measured through the use of indicators, for which data are collected and recorded. The analysis of the indicators then informs us on the level of performance.

Performance indicators are introduced here in the context of their place in the performance assessment framework; greater detail is provided in Chapter 3.

The linkage between the criteria against which performance is to be measured, and the indicators that are to be used to measure attainment of those criteria, is important. Using the nested systems outlined in Fig. 2.3, for example, performance criteria and indicators for the irrigation system, the agricultural system and the agricultural economic systems can be defined (Table 2.7). Note that a performance criterion, such as equity, can be defined differently depending on the system to which it relates.

In some instances it is useful to consider indicators for the inputs and outputs across a number of systems; examples are presented in Table 2.8.

Target values may be set for these indicators, or the values obtained at a particular location or time can be compared with values of the indicator collected at other locations (*spatial variation*) or time (*temporal*

Table 2.6. Comparison of objectives, weightings and rankings for a state farm and a settlement scheme (Burton, 1993).

Objective	State farm[a] Weighting	State farm[a] Ranking	Settlement scheme Weighting	Settlement scheme Ranking	Performance indicator	Target value
Maximize area harvested	6	(v)	10	(ii)	Area harvested	2052 ha (100%)
Maximize total production	10	(iv)	6	(iii)	Total production	7600 t
Maximize total value of agricultural production	10	(i)	6	(iv)	Total value of production	$1,067,238
Maximize productivity of land (kg/m³)	10	(ii)	10	(v)	Water productivity	0.16 kg/m³
Maximize productivity of water ($/m³)	10	(iii)	10	(vi)	Value of production per unit water	$0.023/m³
Maximize equity of water supply	0	(vi)	10	(i)	Area planted/ area harvested	1.0
					Delivery performance ratio	SD < 10%

[a]For weightings 1 is low, 10 is high; for ranking (i) is highest, (vi) is lowest.

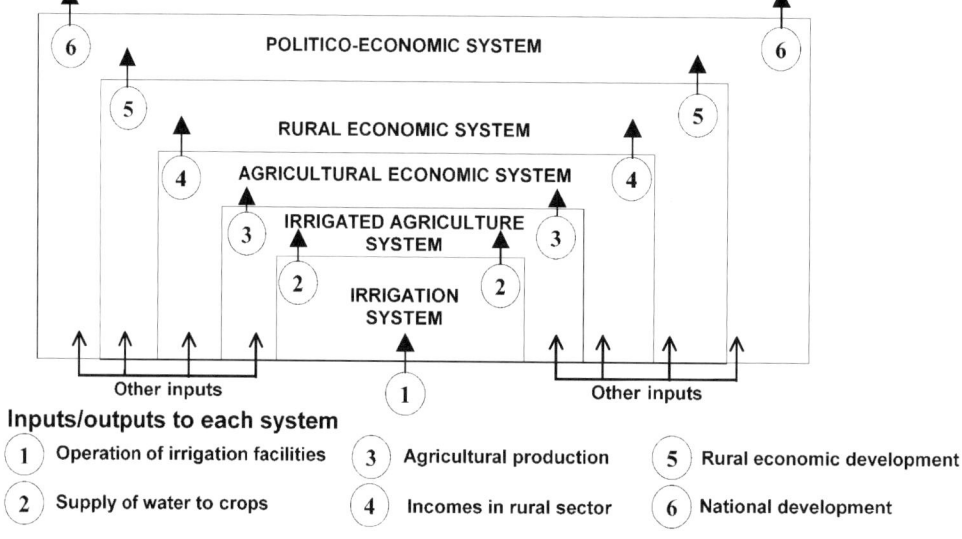

Fig. 2.3. Irrigation in the context of nested systems (Small and Svendsen, 1992).

Table 2.7. Examples of linkages between performance criteria and performance indicators. O&M, operation and maintenance.

	Performance indicator[a]		
Criteria	Irrigation and drainage system[b]	Irrigated agriculture system[b]	Agricultural economic system[b]
Command	Water level ratio	–	–
Adequacy	Overall consumed ratio Delivery performance ratio	Crop production relative to family food needs	Cash value of crop production relative to defined poverty level
Equity	Overall consumed ratio Delivery performance ratio	Spatial distribution within scheme of: – crop type – crop yield – cropping intensity	Spatial distribution within scheme of farm income
Reliability	Overall consumed ratio Delivery performance ratio	Number of years crop production is adequate	Number of years income from crop production is adequate
Efficiency	Overall consumed ratio Field application ratio Outflow over inflow ratio	Crop yield	O&M fraction
Productivity	–	Crop yield	Crop gross margin Internal rate of return
Profitability	–	–	Farm profit Return on investment (EIRR)
Sustainability	Efficacy of infrastructure Groundwater depth Indicator value on salinity	Sustainability of irrigable area	Financial self-sufficiency O&M fraction Fee collection ratio

[a]See Chapter 3 for more detail on some of these indicators.
[b]As detailed in Fig. 2.3.

variation). Thus values of performance indicators can be compared within or between schemes.

Data requirements

Following on from identification of the performance criteria and indicators to be used in the performance assessment programme, the data needs can be identified (Table 2.9).

Data collection (who, how, where and when)

During the design stage of the performance assessment programme it will be necessary to identify: *who* will collect these data, and *how*, *where* and

Table 2.8. Examples of indicators using inputs and outputs across different systems.

Criteria	Indicator example	Systems covered
Productivity	Land productivity (kg/ha)	Irrigation system Irrigated agriculture system
Productivity	Water productivity ($/ha)	Irrigation system Agricultural economic system

Table 2.9. Linking performance indicators to data requirements.

Indicator[a]	Definition	Units	Data required
Cropping intensity	$\dfrac{\text{Actual cropped area}}{\text{Irrigable area}}$	%	Actual cropped area (ha) Irrigable area (ha)
Crop yield	$\dfrac{\text{Crop production}}{\text{Area cultivated}}$	kg/ha	Crop production (kg) Area cultivated (ha)
Overall consumed ratio	$\dfrac{\text{Crop water demand} - \text{Effective precipitation}}{\text{Volume of water supplied to command area}}$	–	Crop water demand (mm) Effective precipitation (mm) Irrigation water supply (mm)
Water productivity	$\dfrac{\text{Yield of harvested crop}}{\text{Volume of supplied irrigation water}}$	kg/m^3	Crop production (kg) Area cultivated (ha) Volume of irrigation water supplied (m^3)

[a] These indicators are defined in more detail in Chapter 3.

when they will be collected. A more detailed account of these procedures is provided in Chapter 6; the explanation here is limited to the context of the performance assessment framework.

All or some of the required data may already be available, such as crop areas, or there may be a need for additional data collection procedures or special equipment to collect data (such as automatic water level recorders to gather detailed information on canal discharges day and night). Allowance will need to be made in the performance assessment budget for the costs associated with the data collection and handling programme.

To understand the performance of an irrigation scheme it is neither necessary, nor economical or time efficient to collect data for every location in a scheme. The performance assessment programme should be designed to take representative samples to enable an adequate analysis to be carried out in keeping with the prescribed needs. It is, for example, common to take sample tertiary units from the head, middle and tail of irrigation systems when studying irrigation water management performance.

When the data needs have been decided, a data collection schedule can then be drawn up. An example schedule for a performance assessment programme by a scheme manager is presented in Table 2.10.

In addition a matrix can be drawn up (Table 2.11) showing the performance indicators to be used and the data to be collected. As can be seen in the example provided, some data apply to a number of indicators.

Form of output

At the planning stage for the performance assessment programme it is helpful to think about the form of the report output. Preparing a draft annotated contents list of the report, and a list of tables and figures and their anticipated content helps focus thinking and ensures that data are collected to match. An example is given (Table 2.12) for a study to gain a broad understanding of performance related to irrigation water supply throughout a scheme.

Simple sketches of the form of the expected output are helpful, as is thinking about the form of data presentation that the users of the performance assessment report and data would find most useful. Non-technical personnel might be interested, for example, in a graph showing the trend in the decline in water quality over time, without requiring too much detail on the actual figures. Technical personnel, however, would require the figures to be presented, perhaps in a table associated with the graph. More details are presented in Chapter 6 on this subject.

Implementation

The performance assessment programme design phase is followed by the implementation phase, covering the actual collection, processing, analysis and reporting of the data. Depending on the nature of the performance assessment programme, implementation may be over a short (1 week) or long (several years) period. In all cases it is worthwhile to process and analyse some, if not all, of the data collected as the work progresses in order to detect errors in data and take corrective action where necessary.

Data collection, processing, analysis and reporting are covered in more detail in Chapter 6.

Application of Output

The use of the information collected from a performance assessment study will vary depending on the purpose of the assessment. The use to which the results of the performance assessment are put will depend on the reason the performance assessment was carried out.

Possible actions following the conclusion of the performance assessment study might include:

1. Redefining strategic objectives and/or targets.
2. Redefining operational objectives and/or targets.

Table 2.10. Example of a data collection schedule – who, how, where and when.

Data required	Units	Who	How	Where	When
Irrigable area	ha	Scheme manager	From design drawings or scheme database	In office	–
Crop production	kg	Scheme agronomist	Interviews with farmers	In selected sample tertiary units	At end of season
Actual cropped area	ha	Scheme agronomist	Data returns from farmers, and/or spot checks in field	For whole scheme but field checks made on selected sample tertiary units	During the irrigation season
Crop yield	kg/ha	Scheme agronomist	Crop cuttings	In selected sample tertiary units	At harvest time
Crop water demand	mm/day	Scheme agronomist or irrigation engineer	By calculation using standard procedures (e.g. CROPWAT or CRIWAR)	In selected sample tertiary units	During the season
Rainfall	mm/day	Water masters	Using rain gauge	At locations within the scheme area	Daily
Actual discharge	m^3/s	Water masters	Reading of measuring structure gauges	At selected sample tertiary unit intakes	Daily
Actual duration of flow	h	Water masters	Reading of measuring structure gauges	At selected sample tertiary unit intakes	Daily
Intended discharge	m^3/s	Scheme manager	From indents submitted by farmers	In office	Each week
Intended duration	h	Scheme manager	From indents submitted by farmers	In office	Each week
Crop market price	$/kg	Scheme agronomist	Interviews with farmers and traders	Villages and markets	At end of season

Note: The example given is for a performance assessment programme carried out by a scheme manager for the whole scheme with a view to understanding overall scheme performance.

Table 2.11. Linking performance indicators to data collection.

Data required	Units	Indicator						
		Cropping intensity (%)	Crop yield (kg/ha)	Overall consumed ratio	Water productivity (kg/m^3)	Delivery performance ratio	Output per unit cropped area ($/ha)	Output per unit irrigation supply ($/m^3)
Irrigable area	ha	✓						
Crop production	kg		✓					
Actual cropped area	ha	✓	✓					
Crop yield	kg/ha		✓		✓		✓	✓
Crop water demand	mm			✓				
Rainfall	mm			✓				
Actual discharge	m^3/s			✓	✓	✓		✓
Actual duration of flow	h				✓	✓		✓
Intended discharge	m^3/s					✓		
Intended duration of flow	h					✓		
Crop market price	$/kg						✓	✓

Note: The example given is for a performance assessment programme carried out by a scheme manager for the whole scheme with a view to understanding overall scheme performance.

Table 2.12. Example of planned figures and tables for a performance assessment programme.

	Content
Figure no.	
1	Layout of irrigation system
2–10	Histogram plots of discharge versus time (daily) at primary, secondary and selected tertiary head regulators
11–16	Histogram plots of irrigation depth applied to a sample number of individual (sample) fields
17–22	Histogram plots of delivery performance ratio for a sample number of individual fields
Table no.	
1	Summary table of performance at head regulator level, including: total command area, irrigated area, total flow (MCM), total days flowing during season, average unit discharge (l/s per ha)
2–6	Summary table of cultivable command area, cropped areas, crop types, cropping intensities for primary and a sample number of secondary and tertiary command areas
7–12	Summary tables of data collected at field level, including, for each sample field: area, crop type, number of irrigations, irrigation depths, irrigation intervals, maximum soil moisture deficit, total water supply, total estimated water demand, crop production and crop market price
13–18	Summary table of results of calculation showing: yield per unit area (kg/ha), yield per unit irrigation supply (kg/m^3), output per cropped area ($/ha), output per unit irrigation supply ($/m^3)

3. Implementing corrective measures, for example:
   - Training of staff.
   - Building new infrastructure.
   - Carrying out intensive maintenance.
   - Developing new scheduling procedures.
   - Changing to alternative irrigation method(s).
   - Rehabilitation of the system.
   - Modernization of the system.
   - Installing field drainage.

Further Action

Further studies may be required as a result of the performance assessment programme. As discussed in Chapter 1 and later in Chapter 5, performance assessment is closely linked with diagnostic analysis. It is often the case that an initial performance assessment programme identifies areas where further measurements and data collection are required in order to identify the root causes of problems and constraints.

Where performance assessment identifies the root cause of a problem or constraint, further studies may be required to implement measures to alleviate the problem: for example, field surveys for the planning and design of a drainage system to relieve waterlogging.

Note

[1] Logical framework: a project planning, monitoring and evaluation tool commonly used by funding agencies to clearly define the project goal, purpose, outputs and actions.

References

Abernethy, C.L. (1984) *Methodologies for Studies of Irrigation Water Management*. Report OD/TN 9. Hydraulics Research, Wallingford, UK.
Bottrall, A.F. (1981) *Comparative Study of the Management and Organisation of Irrigation Projects*. World Bank Staff Working Paper No. 458. World Bank, Washington, DC.
Burt, C.M. and Styles, S.W. (1999) Modern water control and management practices in irrigation – impact on performance. Water Report No. 19, Food and Agriculture Organization of the United Nations, Rome.
Burton, M.A. (1993) A simulation approach to irrigation water management. Unpublished PhD thesis, University of Southampton, UK.
Burton, M.A. and Mututwa, I.M. (2002) A methodology for performance assessment of irrigation and drainage systems (draft paper).
Chambers, R. (1988) *Managing Canal Irrigation: Practical Analysis from South Asia*. Cambridge University Press, Cambridge.
Ijir, T.A. and Burton, M.A. (1998) Performance assessment of the Wurno Irrigation Scheme, Nigeria. *ICID Journal*, 47 (1). International Commission on Irrigation and Drainage, New Delhi.
Molden, D.J. and Gates, T.K. (1990) Performance measures for evaluation of irrigation water delivery systems. *Journal of Irrigation and Drainage Engineering, ASCE*, 116(6).
Murray-Rust, D.H. and Snellen, W.B. (1993) *Irrigation System Performance Assessment and Diagnosis*. Joint IIMI/ILRI/IHEE Publication. International Irrigation Management Institute, Colombo, Sri Lanka.
Mututwa, I.M. (2002) A framework for performance assessment of irrigation and drainage schemes. Unpublished MPhil thesis, University of Southampton, UK.
Oad, R. and McCornick, P.G. (1989) Methodology for assessing the performance of irrigated agriculture. *ICID Bulletin* 38 (1). International Commission on Irrigation and Drainage, New Delhi.
Sagardoy, J.A., Bottrall, A. and Uittenbogaard, G.O. (1982) Organisation, operation and maintenance of irrigation schemes. Irrigation and Drainage Paper No. 40, Food and Agricultural Organization of the United Nations, Rome.
Small, L.E. and Svendsen, M. (1992) *A Framework for Assessing Irrigation Performance*. IFPRI Working Papers on Irrigation Performance No. 1. International Food Policy Research Institute, Washington, DC, August.
Svendsen, M. (1990) Choosing a perspective for assessing irrigation system performance. Paper presented at the FAO Regional Workshop on Improved Irrigation System Performance for Sustainable Agriculture, Bangkok, Thailand, 22–26 October.

3 Performance Indicators for Irrigation and Drainage

Characteristics and Application of Performance Indicators

As follows from its definition (Chapter 1), performance assessment is a tool: (i) to improve the level of service or operation between irrigation-related institutions; and (ii) to improve the efficiency with which resources are being used.

It is important to ensure that indicators that are selected to quantify the performance for a system describe performance in respect to the objectives established for that system. A meaningful indicator can be used in two distinct ways. It tells a manager what the current performance is of the system and, in conjunction with other indicators, may help him to identify the correct course of action to improve performance within that system. In this sense the use of the same indicator over time is important because it assists in identifying trends that may need to be reverted before the remedial measures become too expensive or too complex. A fuller description of desirable attributes of performance indicators is given in Table 3.1.

As mentioned above, the ultimate purpose of performance assessment is to achieve efficient, productive and effective irrigation and drainage systems by providing relevant feedback to management at all levels. As such, it may assist management or policy makers in determining whether performance is satisfactory and, if not, which corrective actions need to be taken in order to remedy the situation.

To determine the related degree of satisfaction, a systematic and timely flow of actual (measured or collected) data on key parameters of a system must be compared with intended or limiting (critical) values of these data. This comparison can be done in two ways:

Table 3.1. Properties of performance indicators (Bos *et al.*, 1994a).

Scientific basis
The indicator should be based on an empirically quantified, statistically tested causal model of that part of the irrigation process it describes. Discrepancies between the empirical and theoretical basis of the indicator must be explicit, i.e. it must not be hidden by the format of the indicator. To facilitate international comparison of performance assessment studies, indicators should be formatted identically or analogously as much as possible (ICID, 1978; Bos and Nugteren, 1990; Wolters, 1992).

The indicator must be quantifiable
The data needed to quantify the indicator must be available or obtainable (measurable) with available technology. The measurement must be reproducible.

Reference to a critical or intended value
This is, of course, obvious from the definition of a performance indicator. It implies that relevance and appropriateness of the critical or intended values and tolerances can be established for the indicator. These values (and their allowable range of deviation) should be related to the level of technology and management (Bos *et al.*, 1991).

Provide information without bias
Ideally, performance indicators should not be formulated from a narrow ethical or disciplinary perspective. This is, in reality, extremely difficult as even technical measures contain value judgements (Small, 1992).

Provide information on reversible and manageable processes
This requirement for a performance indicator is particularly sensible from the irrigation manager's point of view. Some irreversible and unmanageable processes could provide useful indicators, although their predictive meaning may only be indirect. For example, the frequency and depth of rainfall are not manageable, but information from a long time series of data may be useful in planning to avoid water shortage and information on specific rainfall events may allow the manager to change water delivery plans.

Nature of the indicator
An important factor influencing the selection of an indicator has to do with its nature: the indicator may describe one specific activity or may describe the aggregate or transformation of a group of underlying activities. Indicators ideally provide information on an actual activity relative to a certain critical or intended value. The possibility of combining such dimensionless ratios into aggregate indicators should be studied, in much the same way that many indicators used for national economic performance are composites.

Ease of use and understanding, and cost-effectiveness
Particularly for routine management, performance indicators should be technically feasible, and easily used by management staff given their level of skill and motivation. Further, the cost of using indicators in terms of finances, equipment and commitment of human resources, should be well within the management's resources.

1. Present the (measured or collected) data through a (dimensionless) performance indicator, whose ratio includes both an actual value and an intended (or critical) value of data on the considered key parameter (Fig. 3.1). As mentioned in Chapter 2, the indicator should have a target level that is based on the 'service agreement'. Around the target level is an

allowable range (either to one or two sides) within which the indicator can fluctuate without triggering a management action. However, if the indicator moves out of this range, diagnosis of the problem should lead to the planning of corrective action.

2. Present the (measured or collected) data and compare the 'measurable parameter' with an intended (or critical) value of this measurable key parameter (Fig. 3.2). In Fig. 3.2 the parameter is plotted as a function of time and with reference to the 'critical level' of this parameter and the related critical deviation. The entering of the parameter within the critical deviation range then triggers the diagnostic management activities.

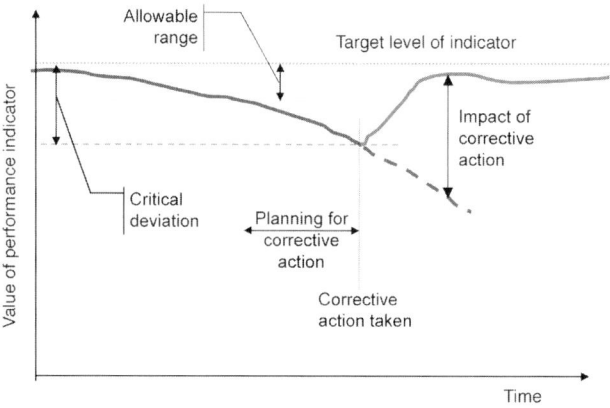

Fig. 3.1. Terminology on the use of a dimensionless performance indicator.

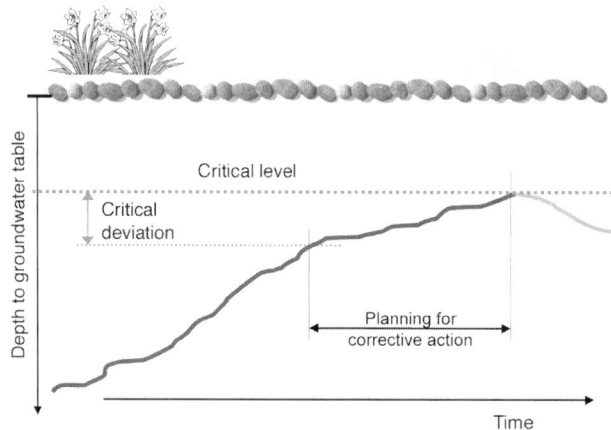

Fig. 3.2. Change in time of a parameter (groundwater depth) and its comparison with the related critical level (to avoid salinization).

Besides a presentation in time, both types of indicators also can be analysed with respect to their spatial distribution.

Types of Performance Indicators

As discussed above, the characteristic activity during performance assessment is the comparison of the measured value of a parameter with the target value, or intended value, of this parameter. In doing so, the terminology in Table 3.2 is proposed.

It is recommended to compare the above values through a dimensionless ratio with the actual (measurable) value of the parameter (of irrigation and drainage) in the numerator. The parameter value in the denominator of the ratio can be divided into four main groups:

Table 3.2. Terminology.

Terminology	Definition	Remarks
Actual value	Something (parameter) that can be measured or determined.	E.g. measured flow rate, crop yield, irrigation fee, groundwater depth.
Benchmark	The desired value of process output parameter (or of the performance indicator).	The benchmark level is set by comparison with best practices of comparable processes.
Critical value	The critical value of the key parameter quantifies a physical process whereby the concentration of a chemical limits crop yield, or hampers health, if a critical value is passed.	E.g. the salinity of irrigation water has a critical value that reduces crop yield if passed.
Intended value	Value of the measurable parameter that the service-providing organization is trying (intends) to achieve.	The intended value should be based on the (agreed) service level or on the strategy.
Key parameter	A quantifiable (measurable) parameter that influences irrigation or drainage performance.	E.g. flow rate, crop yield, irrigation fee, groundwater depth.
Service level	Amount of goods or services (e.g. water) provided by a service provider to a user. The user can be another organization, a person or group of people, deemed necessary for proper and effective functioning.	Should be based on the (national) water law and a service arrangement or agreement between providers and users.
Target value	The desired value of a performance indicator.	See also benchmark.
Total value	The total number (or the sum) of a parameter.	E.g. number of water users, number of structures, etc.

1. The *critical value* of a key parameter is used if the assessed process is physically determined or shows a similar behaviour. Commonly, these indicators describe one specific parameter. Most of the indicators in this group can be (or are) used in strategic performance assessment.
2. The *intended value* of the key parameter is used if a human decision is involved in setting this value. The indicator often describes the aggregate or transformation of a group of underlying activities. Most of the indicators in this group can be (or are) used in operational performance assessment.
3. The *(actual) input value* of the key parameter is used to quantify the output over input ratios of key resources. This group of ratios resembles the classical efficiencies of water use, etc.
4. The *total value* of the key parameter is used to quantify the actual performing fraction (percentage) of a total available resource. Most of these indicators relate to socio-economic (budgetary) parameters of irrigation management.

Although we recommend the use of dimensionless indicators, we do not intend to discourage the use of the 'measurable' value of a parameter. In particular, the presentation of a (measurable) parameter as a function of time, being supplemented with the related critical value, gives clear information on performance with respect to this parameter (Fig. 3.2). Examples of such presentations will be given at selected places in this chapter.

Selected Performance Indicators

Taking the properties of Table 3.1 into account, indicators were selected and defined to assess the performance of water management (in irrigation and drainage). Table 3.3 lists performance indicators that are recommended for general use. The indicators are grouped into four categories:

1. Water balance, water service and maintenance. The indicators in this group refer to the primary function of irrigation and drainage; the provision of a water service to users.
2. Environment. Both irrigation and drainage are man-made interventions in the environment to facilitate the growth of crops. The non-intentional (mostly negative) effects of this intervention are considered in this group.
3. Economics. This group contains indicators that quantify crop yield and the related funds (generated) to manage the system.
4. Emerging indicators. This group gives four indicators that contain parameters which need to be measured by use of satellite remote sensing. This emerging technology enables very cost-effective measurement of data.

As mentioned in Chapter 2, the number of indicators needed for an assessment depends on boundary conditions and on the purpose of the

Table 3.3. Four groups of performance indicators.

Performance indicator value	Type of assessment
$\dfrac{\text{Actual value of key parameter}}{\text{Critical value of key parameter}}$	Actual physical processes whereby a critical value limits either crop yield or the sustainability of agriculture in the considered area.
$\dfrac{\text{Actual value of key parameter}}{\text{Intended value of key parameter}}$	Classical comparison of an actual physical situation with respect to an intended value. Most indicators relate to water delivery.
$\dfrac{\text{Actual output value of key parameter}}{\text{(Actual) input value of key parameter}}$	Assessment of the efficiency with which a resource (water, land, funds, etc.) is used. The classical irrigation efficiencies fall in this group.
$\dfrac{\text{Actual value of key parameter}}{\text{Total value of key parameter}}$	Assessment of the fraction (percentage) of infrastructure (resource) that functions.

assessment. It is recommended that performance is assessed from different perspectives. Thus, indicators from each of the above four groups should be combined in the assessment programme. To minimize the cost of data measurement, however, it is recommended to justify the use of each selected indicator. If the recommended list does not meet all demands, additional indicators may be selected from Appendices 1 and 2. As with all data, one single indicator does not give sufficient information to support a (management) decision. The added information from data is obtained if the indicator values are presented either:

- As a function of time. In this way the trend of the indicator with respect to its target level can be studied.
- With respect to its spatial distribution. In this way the indicator values of different units (command areas) within the same irrigated (or drained) area can be compared and correlated with other parameters.

Definition of Performance Indicators

This section defines the indicators as listed in Table 3.4. The potential use of each indicator is illustrated with one example.

Water balance, water service and maintenance

The water-related indicators focus on the 'core business' of irrigation: the diversion and conveyance of water to individuals or groups of users or to other sectors. These indicators are concerned with how well water supply matches demand, whether services are reliable, adequate and timely and whether social equity has been met.

Table 3.4. Selected performance indicators with their function if used to quantify a trend in time or a spatial distribution.

Relationship	Information provided if the indicator is used to show:	
	A trend in time	The spatial distribution
Water Balance, Water Service, and Maintenance		
Overall consumed ratio	Degree to which irrigation water requirements of the users (farmers, urban users, industry, environment, etc.) were met	Shows difference in water supply to users at various locations within command area. Quantifies the uniformity and equity of water supply
Field application ratio	Changes of water use by irrigators	Influence of different boundary conditions on (efficiency of) irrigation water use
Depleted fraction	Show changes in actual water use	Quantifies differences in the water balance of considered (command) areas
Drainage ratio	Degree to which water within the drainage basin is consumed	Identifies areas where water resources can be developed
Outflow over inflow ratios	Quantifies the need for maintenance of system components	For identification of system components (physical or organizational) that need maintenance, improvement or modernization
Delivery performance ratio	Shows changes in quality of service to water users	Quantifies the uniformity and equity of water delivery
Dependability of interval between water applications	Shows changes in service (timing only) to water users	Illustrates the equity (timing only) of service (water delivery) to water users
Canal water level and head–discharge relationship	Quantifies the need for maintenance of system components	Identifies system components that need repair or replacement
Effectivity of infrastructure	Quantifies effect of maintenance	Shows areas with maintenance problems
Environment		
Groundwater depth	Indicates problems of rising or falling groundwater levels	Shows areas with potential waterlogging or salinity problem, or areas where groundwater is mined
Pollution of water	Indicates pollution level in relation to critical value	Shows areas with potential water pollution

Table 3.4. *Continued.*

Relationship	Information provided if the indicator is used to show:	
	A trend in time	The spatial distribution
Sustainability of irrigable area	Quantifies the intensity of land occupancy by crops in the irrigated area	Quantifies crop occupancy rate of sub-areas
Economics		
Water productivity	Quantifies change in crop yield or value per m³ water supplied	Shows spatial variation in productivity (kg/m³)
Land productivity	Quantifies change in crop yield or value per unit area	Shows spatial variation in productivity (kg/ha)
MO&M funding ratio	Shows changes in the financial viability of the water management institution	Quantifies relative performance of management units within a system (e.g. WUAs)
O&M fraction	Quantifies the adequacy of funds for O&M tasks	Quantifies relative performance of management units within a system (e.g. WUAs)
Fee collection ratio	Shows degree to which water users (are willing to) pay for the water delivery service	Detects areas where water users do not pay
Relative water cost	Shows changes in water cost	Shows areas where water is relatively more expensive to obtain
Price ratio	Shows changes in marketing conditions for irrigated crop(s)	Shows areas where farmers may have to stop irrigation
Emerging indicators		
Crop water deficit	Quantifies reduction in evapotranspiration	Detects water-short areas
Relative evapotranspiration	Quantifies relative reduction in evapotranspiration	Detects water-short areas
Relative soil wetness	Quantifies changes in the availability of water for crop growth	Shows areas with water shortage, drainage problems, etc.
Biomass production per m³ water supply	Quantifies change in biomass production per m³ water supplied	Shows spatial variation in production of biomass (kg/m³)

Overall consumed ratio

The overall (or project) consumed ratio (efficiency) quantifies the degree to which the crop irrigation requirements are met by irrigation water in the irrigated area (Bos and Nugteren, 1974; Willardson et al., 1994). Assuming negligible non-irrigation water deliveries to the area, the ratio is defined as (Bos and Nugteren, 1974[1]):

$$\text{Overall consumed ratio} = \frac{ET_p - P_e}{\text{Volume of water supplied to command area}}$$

where ET_p = potential evapotranspiration, P_e = effective precipitation.

The numerator of this indicator originally (ICID, 1978) contains: 'the volume of irrigation water needed, and made available, to avoid undesirable stress in the crops throughout (considered part of) the growing cycle'. This value of $(ET_p - P_e)$ for the irrigated area is entirely determined by the crop, the climate and the interval between water applications. Hence, the actual value of the overall consumed ratio varies with the actual volume of irrigation water supplied to the considered command area. The value of $(ET_p - P_e)$ can be calculated by use of models like CRIWAR (Bos et al., 1996) and CROPWAT (Smith et al., 1991). Because the total water supply to a command area (irrigation project) is among the very first values that should be measured (together with the cropped area, the cropping pattern and meteorological data), the overall consumed ratio is the first indicator that should be available for each irrigated area. An example of three years of monthly ratios is given in Fig. 3.3.

Figure 3.3 quantifies the effect of a number of water management practices. As will be mentioned, some of these practices have undesirable side effects.

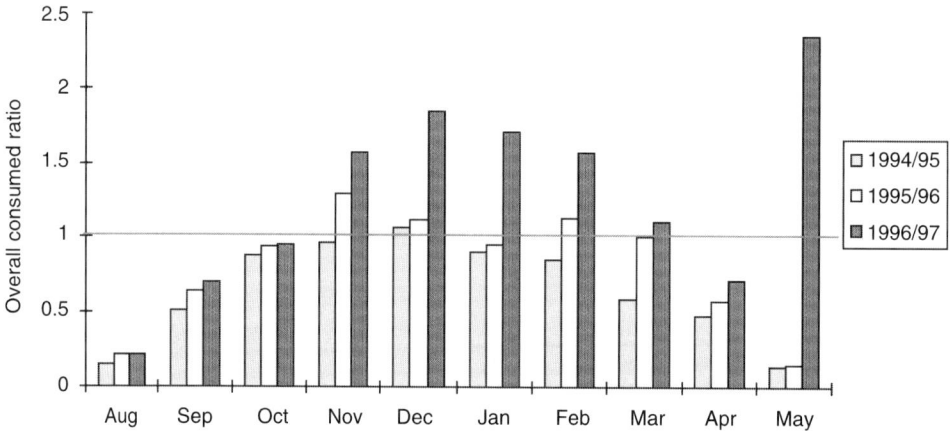

Fig. 3.3. The overall consumed ratio as a function of time for three irrigation seasons (Tunayan, Argentina) (Morabito et al., 1998).

- During a season with sufficient water supply (1994/95), the overall consumed ratio varies from very low (< 0.2) at the start and end of the irrigation season, to a ratio above 1 during the peak month(s).
- During periods with low ratios, the non-consumed fraction of the water will cause the groundwater table to rise (only if this water is applied to the field), while during periods with a ratio above 0.6 groundwater must be pumped and stored to avoid water shortage (Bos et al., 1991).
- With less water becoming available (1995/96), the number of months with a ratio over 1 increased. If water were managed in such a way that the ratio varies between 0.5 and 0.6 during the off-peak months, allowing water to be stored, the period with water shortage could be shortened considerably.
- During the very dry 1996/97 season, the water supply to the users' associations (UAs) was reduced by 30%. This reduction was decided upon (thus again a water management decision) following the October forecast of available water recourses. In early May the canals had to be closed because the storage reservoir was empty.

The overall consumed ratio also can be quantified for each lateral or tertiary unit and presented with a spatial distribution. Within an existing irrigated area we recommend setting a target overall consumed ratio, and compare the actual ratio at a monthly and annual basis with this target value.

Field application ratio

The field application ratio (efficiency) has the same structure as the overall consumed ratio. It is defined as (ICID, 1978):

$$\text{Field application ratio} = \frac{ET_p - P_e}{\text{Volume of water delivered to field(s)}}$$

The numerator of this indicator originally contains: 'the volume of irrigation water needed, and made available, to avoid undesirable stress in the crops throughout (considered part of) the growing cycle'. This 'volume' is expressed in terms of m³/ha or in terms of water depth. The numerator equals the potential evapotranspiration by the irrigated crop minus the effective part of the precipitation: $ET_p - P_e$.

The value of $(ET_p - P_e)$ is entirely determined by the crop, the climate and the interval between water applications. Hence, the value of the field application ratio varies with the actual volume of irrigation water delivered to the field. This water delivery depends on the reliability of the 'service' by the water-providing agency, the irrigation know-how of the farmer and the uniformity with which water can be applied to the field (thus on the water application technology). From a technology point of view, attainable values of the field application ratio (efficiency) are shown in Table 3.5. These in essence provide benchmark values against which targets can be set.

The calculation period of the field application ratio depends on the (average) interval between water applications to the fields. If the period is too short, the number of water applications varies per period. It is recommended to use a calculation period that contains at least two water applications. One month is a suitable minimal period.

In arid and semi-arid areas the field application ratio with a calculation period of one irrigation season should remain below 0.90 to avoid salt accumulation in the root zone of the irrigated crop. Hence, from a sustainability point of view it does not make sense to try to be 'too efficient' in irrigation water use. Therefore, the target value is below the maximum attainable value of Table 3.5.

Depleted fraction

The depleted fraction is the ratio that compares three components of the water balance of an irrigated area. This indicator is particularly useful for diagnostic purposes in water-scarce areas. The depleted fraction relates the actual evapotranspiration from the selected area to the sum of all precipitation on this area plus surface water inflows into the irrigated area (typically irrigation water). It is defined as (Molden, 1998; Molden and Sakthivadivel, 1999):

Table 3.5. Common maximum attainable values of the field application ratio (efficiency) (Bos, 1974, 1982; Jurriens *et al.*, 2001).

Irrigation water application method	Maximum attainable ratio (efficiency)
Surface irrigation	
Furrows, laser levelling	0.70
other quality levelling methods	0.60
Border strip, laser levelling	0.70
other quality levelling methods	0.60
Level basins, laser levelling	0.92
other quality levelling methods	0.80
Sprinkler	
Hand move system	0.60
Overhead rain drops	0.80
Downward fine spray	0.90
Micro-irrigation	
Drip	0.95
Micro sprinkler	0.95

$$\text{Depleted fraction} = \frac{ET_a}{P_e + V_c}$$

where ET_a = actual evapotranspiration from the gross command area; P_e = precipitation on the gross command area; V_c = volume of surface water flowing into the command area.

Because it is not practical to measure the ET_a and the precipitation for only the irrigated part of the area, we consider the gross command area. As shown in Fig. 3.4, the depleted fraction quantifies the surface water balance excluding the drainage component. The water manager can influence the value of V_c, while this in turn influences the water deficit $(ET_p - ET_a)$ in the area.

Due to the above definition of the water balance components, the depleted fraction usually is quantified for the entire irrigated area. We recommend studying the depleted fraction as a function of time. Figure 3.5 shows monthly values for a gross area of 33,800 ha. For semi-arid and arid regions the 'critical value' of the depleted fraction ranges between 0.5 and 0.7 (average about 0.6) (Bastiaanssen et al., 2001; see also Chapter 6).

A critical value of $DP = 0.6$ implies that if ET_a is less than about $0.6(P+V_c)$, a portion of this available water goes into storage, causing the groundwater table to rise, while storage decreases if ET_a is greater than $0.6(P+V_c)$. Thus, the depleted fraction can be used as a performance indicator in irrigation water use. The volume of water diverted into the irrigated area can be reduced during months with a low depleted fraction. If

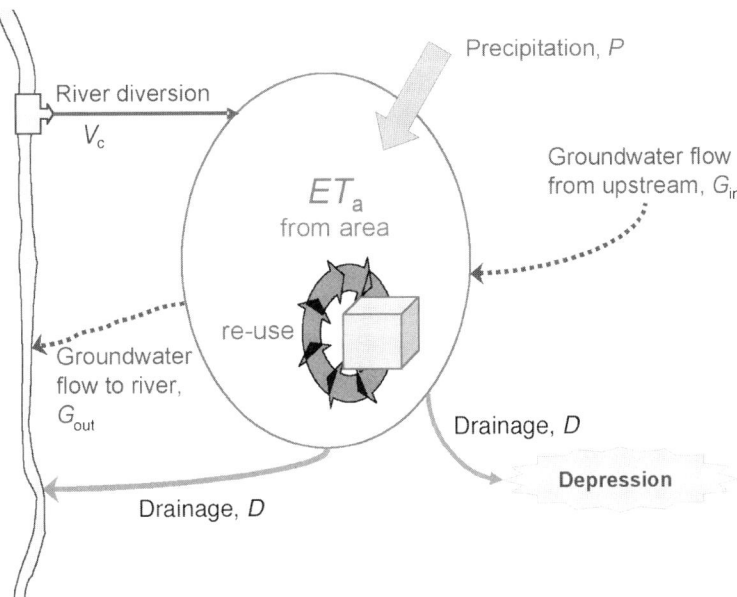

Fig. 3.4. Schematic representation of flows in the water balance of an irrigated area.

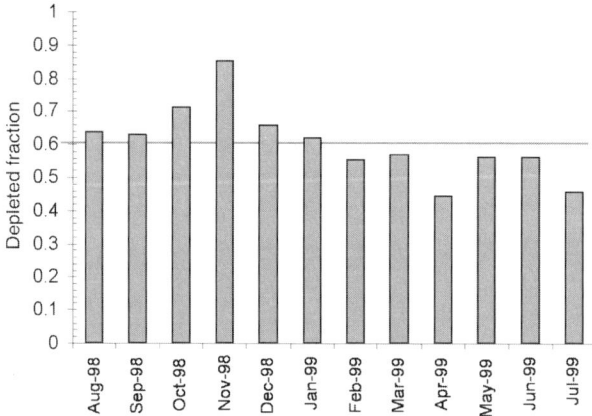

Fig. 3.5. The depleted fraction for the Nilo Coelho project, Brazil (Bastiaanssen *et al.*, 2001).

this non-diverted water remains in a storage reservoir, which often is the case in arid and semi-arid regions, this water can be diverted during dry months.

Drainage ratio

With the increasing scarcity of water, particularly in arid and semi-arid regions, the question of the quantity (volume per month or year) of water that is available for new water users becomes increasingly significant. This question can be posed at different scales, e.g. river basin system, tributary, drainage system, and can be quantified by the drainage ratio that is defined as (Bos *et al.*, 1994b):

$$\text{Drainage ratio} = \frac{\text{Total drained water from area}}{\text{Total water entering into the area}}$$

The drainage ratio is intended to quantify water use in (part of) a river basin with well-defined boundaries. To illustrate the use of this ratio, Table 3.6 gives annual values for three basins. If a value of 0.15 is considered as the critical lower limit to avoid salt accumulation in the drained area, it is obvious that there is little free water for new users in all three river basins.

Considering the water balance of a river basin ($G_{in} = 0$ and G_{out} is relatively small), the drainage ratio is equal to about (1 − depleted fraction).

Outflow over inflow ratios

The classical ratios used to quantify the water balance of a canal system (or reach) are the 'outflow over inflow ratios' (often named efficiency). All ratios have the same structure, being:

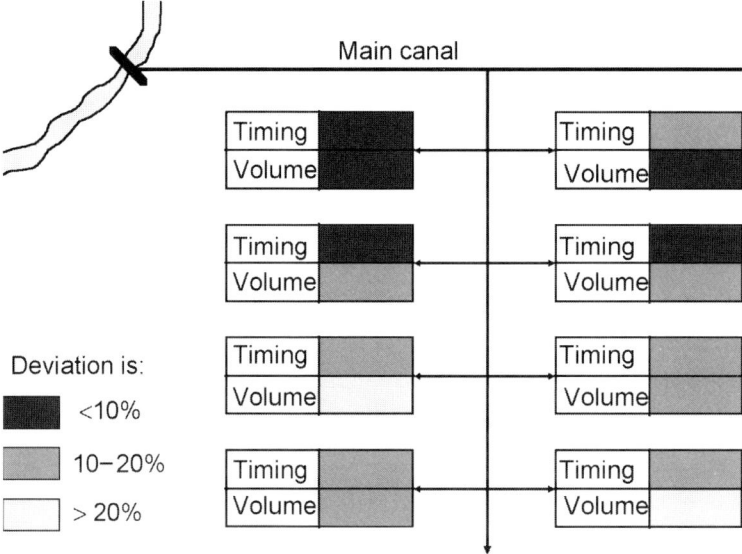

Fig. 3.9. Example of the spatial presentation of two formats of the delivery performance ratio (DPR) in terms of 'time' and 'water volume' for each tertiary unit.

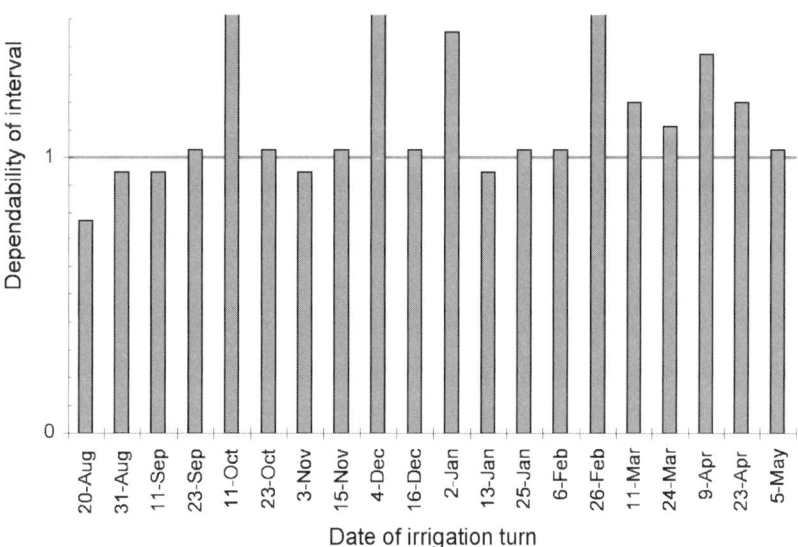

Fig. 3.10. Dependability of the irrigation interval for the Los Sauces Unit (128 ha), Mendoza (Bos *et al.*, 2001).

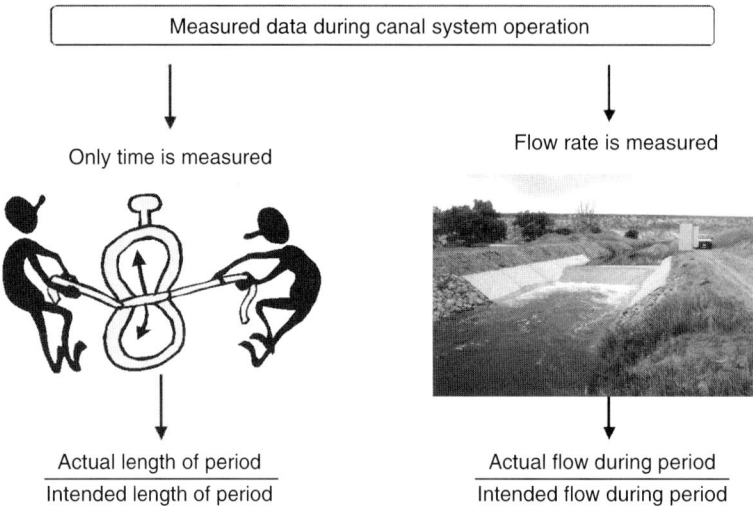

Fig. 3.8. Depending on the available data, the delivery performance ratio will have different formats.

tures with water level recorders must be available at key water delivery locations (Bos, 1976). To facilitate the handling of data, recorders that write data on a chip are recommended (Clemmens *et al.*, 2001).

Over a sufficiently long timescale (e.g. monthly, or over three or four rotational time periods), it can be assumed that, if the delivery performance ratio is close to unity (being the target value), then the management inputs must be effective. Thus, if in Fig. 3.9 all blocks are black, the actual water delivery is as intended. Uniformity of water delivery is high if all units have the same colour. For example, if there is water shortage, and all units are light grey, water management (operation) is such that the burden is equal to all units. Uniformity of water delivery can be quantified by the standard deviation of all DPR-values in the command area.

Dependability of irrigation interval between water applications

The pattern in which water is delivered over time is directly related to the overall consumed ratio of the delivered water, and hence has a direct impact on crop production. The rationale for this is that water users apply more irrigation water if there is an unpredictable variation in timing of delivered water. Also, they may not use other inputs such as fertilizer in optimal quantities if they are more concerned with crop survival (because water is not delivered) than crop production.

The primary indicator proposed for use in measuring dependability of water delivery is concerned with the time between deliveries compared with the plan or subscription. Dependability is defined as:

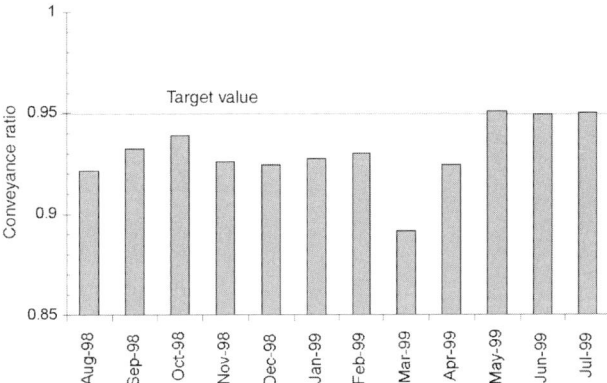

Fig. 3.7. Monthly values of the conveyance (outflow over inflow) ratio of the Nilo Coelho main canal (concrete lined, design capacity 20 m³/s, length 32.5 km).

Delivery performance ratio

The simplest, and yet probably the most important, operational performance indicator is the delivery performance ratio (DPR) (Clemmens and Dedrick, 1984; Clemmens and Bos, 1990; Molden and Gates, 1990; Bos *et al.*, 1991). In its basic form, it is defined as:

$$\text{Delivery performance ratio} = \frac{\text{Actual flow of water}}{\text{Intended flow of water}}$$

Depending on the availability of data, the above 'flow of water' can be determined in two ways (Fig. 3.8):

1. In systems where no structures are available to measure the flow rate, time is the only remaining parameter to quantify water delivery performance. As shown in Fig. 3.8, the DPR then compares the actual length of the water delivery period with intended period. For operational purposes it is then assumed that the flow rate is constant during a relatively long period.
2. With systems dependent on flow rates and volumes, flow rates must be measured (in m³/s). Delivery performance of water then relates the actual delivered volume of water with respect to the intended volume. The length of the period for which the volume is calculated depends on the process that needs to be assessed. It varies from 1 s (for flow rate), one irrigation rotation (for water availability) to 1 month or year (for water balance studies).

The delivery performance ratio enables a manager to determine the extent to which water is actually delivered as intended during a selected period and at any location in the system. It is obvious that if the actually delivered volume of water is based on frequent flow measurements, the greater the likelihood that managers can match actual to intended flows. To obtain sufficiently accurate flow data, discharge measurement struc-

Table 3.6. Annual values of the drainage ratio (Bos and van Aart, 1996).

Drained area (river basin)	Drainage ratio
Aral Sea basin	0.17
Nile in Egypt	0.21
Indus (Pakistan)	0.22

$$\text{Outflow over inflow ratio} = \frac{\text{Total water supply from canal}}{\text{Total water diverted or pumped into the canal}}$$

For large irrigation systems it is common to split the outflow over inflow ratio over different management units of the system. In this context we recommend considering: (i) the conveyance ratio of the upstream part of the system as managed by the irrigation authority; and (ii) the distribution ratio of the WUA-managed canal system. Figure 3.6, for example, illustrates an irrigation canal system with only one source of surface water (no groundwater is pumped into the canal) that supplies water to a number of lateral canals. The conveyance ratio of the main canal then equals $V_c / \Sigma V_{sec,i}$. The ratio should be calculated over a short (month) and a long (season) period (Fig. 3.7).

The rate of change of the ratio is an indicator for the need of maintenance, for example. Quantifying the outflow over inflow ratio for only 1 month gives information to the system manager, provided a target value of the ratio is known. A regular repetition of the measurement allows the assessment of the trend of an indicator in time. This assists the manager in identifying trends that may need to be reversed before the remedial measures become too expensive or too complex.

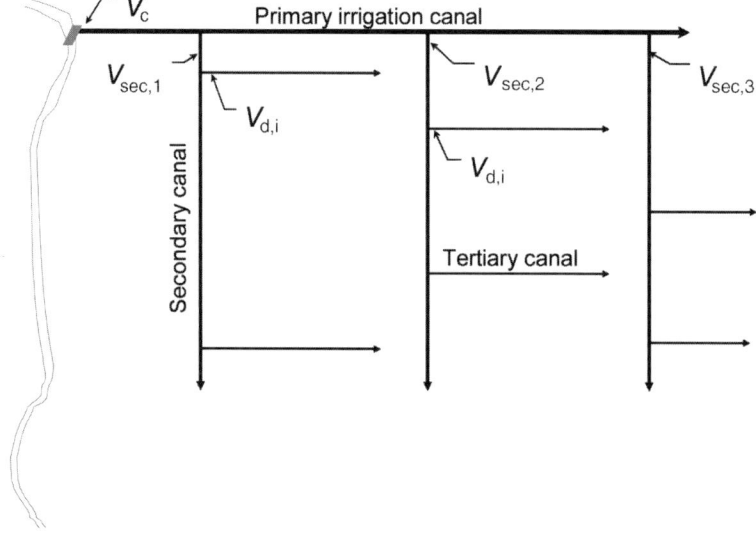

Fig. 3.6. Schematic of an irrigation canal system.

$$\text{Dependability of irrigation interval} = \frac{\text{Actual irrigation interval}}{\text{Intended irrigation interval}}$$

The irrigation interval is measured as the time between the beginnings of two successive water applications. The ditch rider opens the gate that delivers water to the irrigation unit (operation). The intended timing follows from the rotational schedule. Figure 3.10 shows the dependability of all irrigation turns during one irrigation season to the Los Sauces unit.

Canal water level and head–discharge relationship

Maintenance of irrigation and drainage systems intends to accomplish the following main purposes:

- Assure safety related to failure of infrastructure, keep canals in sufficiently good (operational) condition to minimize seepage or clogging, and sustain canal water levels and designed head–discharge relationships.
- Keep water control infrastructure in working condition.

In irrigation systems the assessment of the change in time of the outflow over inflow ratio of the conveyance system provides the best way of assessing whether (canal) maintenance is required. By tracking the change in the ratio over time, it should be possible to establish criteria that will indicate when canal cleaning or reshaping is necessary (Fig. 3.11).

During the design of a canal system, a design discharge and related water level is determined for each canal reach. The hydraulic performance of a canal system depends greatly on the degree to which these design values are maintained. For example, higher water levels increase seepage and cause danger of overtopping of the embankment. Both lower and higher water levels alter the intended division of water at canal bifurcation structures. The magnitude of this alteration of the water

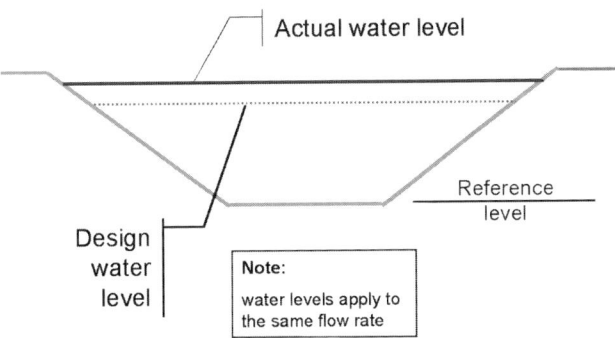

Fig. 3.11. Illustration of terminology.

distribution depends on the hydraulic flexibility of the division structures (Bos, 1976). This change of head (water level) over structures in irrigation canals is the single most important factor disrupting the intended delivery of irrigation water (Bos, 1976; Murray-Rust and van der Velde, 1994).

An indicator that gives practical information on the sustainability of the intended water level (or head) is:

$$\text{Water level ratio} = \frac{\text{Actual water level}}{\text{Design water level}}$$

For closed irrigation and drainage pipes (visual) inspection of heads (pressure levels) is complicated. The functioning of a pipe, however, should be quantified by the measured discharge under a measured head-differential between the upstream and downstream ends of the considered pipe (as used in the original design), versus the theoretical discharge under the same head differential. Hence, pipe performance can be quantified by the ratio:

$$\text{Discharge capacity ratio} = \frac{\text{Actual discharge capacity}}{\text{Design discharge capacity}}$$

The same discharge capacity ratio can be used to quantify the effective functioning of flow control structures in the canal system. Depending on the type of structure, the actual discharge then must be measured under the same (design) differential head (submerged gates, culverts, etc.) or under the same upstream sill-referenced head (free flowing gates, weirs, flumes, etc.). Generally, a deviation of more than 5% would signal the need for maintenance or rehabilitation for flow control structures. Table 3.7 gives an example of the effect of design and construction quality on the performance of subsurface drains.

As mentioned above, maintenance is needed to keep the system in operational condition. For this to occur, (control) structures and water application systems must be operational as intended. Data from the above two ratios can be summarized to quantify maintenance performance by the following ratio:

$$\text{Effectivity of infrastructure} = \frac{\text{Functioning part of infrastructure}}{\text{Total infrastructure}}$$

Table 3.7. Values of the discharge capacity ratio for subsurface drains discharging into concrete or plastic collectors (Shereishra Pilot Area) (Bos et al., 1994a).

	Q_{actual} / Q_{design}	
	Spring	Summer
Into concrete collector pipe	0.31	0.40
Into plastic collector pipe	0.17	0.30

The above three ratios indicate the extent to which the system manager is able to control water. For the analysis to be effective, however, structures should be grouped according to their hierarchical importance (primary, secondary, tertiary and quaternary) and the analysis completed for each level.

Environment

Irrigation can be considered as a human intervention in the environment; water is imported into an area to grow a crop that would not grow without this imported water. In reverse, drainage discharges water from an area to improve crop growth, accessibility of fields, discharge salts from the area, etc. Besides the intended impacts, there are unintended impacts (usually labelled negative, but can be positive). The intended impacts are mostly restricted to the irrigated (or drained) area, while the unintended impacts may spread over the irrigated area, the river basin downstream of the water diversion and the drainage basin downstream of the drained area.

Groundwater depth

Many of the adverse environmental impacts of irrigation are related to the rate of change of the depth to the groundwater table.

- Because of ineffective drainage, or delay in constructing drainage systems in comparison to the surface water supply infrastructure, the groundwater table often rises into the root zone of the irrigated crop. In arid and semi-arid regions this often leads to the increase of capillary rise over seepage, resulting in salinity in the root zone.
- If groundwater being pumped for irrigation exceeds the recharge of the aquifer, the groundwater table drops. As a result, energy cost for pumping may increase to such a level that water becomes too expensive, or groundwater mining may deplete the resource.

For waterlogging and salinity, the critical groundwater depth mostly depends on the (effective rooting depth) of the crop, the overall consumed ratio of irrigation water use and the hydraulic characteristics of the (unsaturated) soil. Depending on these conditions, the critical depth varies between 0.5 and 4 m.

In the case of groundwater mining, the critical depth depends on the cost of pumping water, the value of the irrigated crop and on the depth of the aquifer. If the actual groundwater depth is near the critical depth, the time interval between readings of the ratio should be near 1 month. One year is suitable for most other purposes.

As mentioned before (Chapter 3), indicators that compare a parameter with a critical value of this parameter can be presented in a graph of the measured parameter (*y*-axis) against time (*x*-axis). The critical value of the parameter then is shown as a line (or band) parallel to the x-axis. Figure 3.12 illustrates a case with a critical groundwater depth of 1 m.

Pollution of water

Within the context of the man-made pollution of water, we distinguish between the *consumption* and the *use* of water.

- If water is *consumed* (by the crop) or *depleted*,[2] it leaves the considered part of the system, and cannot be consumed or reused in another part of the considered system. For example, if the field application ratio (efficiency) for a considered field is 55%, this means that 55% of the applied water is evapotranspirated and that the other 45% either becomes surface run-off or recharges the aquifer. Part of this 45% may have been *used* to serve other purposes, e.g. simplify farm management, leaching, etc.
- During the irrigation process water can be *used* for a variety of non-consumptive purposes. These may be related directly with irrigation (facilitate management, silt flushing, leaching, seepage, etc.), or be related with other user groups (energy production, shipping, urban and industrial use, recreation, etc.). As a general rule we may assume that the quality of water decreases upon its use. The indicators in this section quantify the effect of user activities on water quality.

The indicators in this section quantify physical processes whereby the concentration of a chemical limits crop yield, or hampers health, if a critical value is passed. The shape of the indicator is:

$$\text{Indicator value of pollution} = \frac{\text{Actual concentration of pollution}}{\text{Critical concentration of pollution}}$$

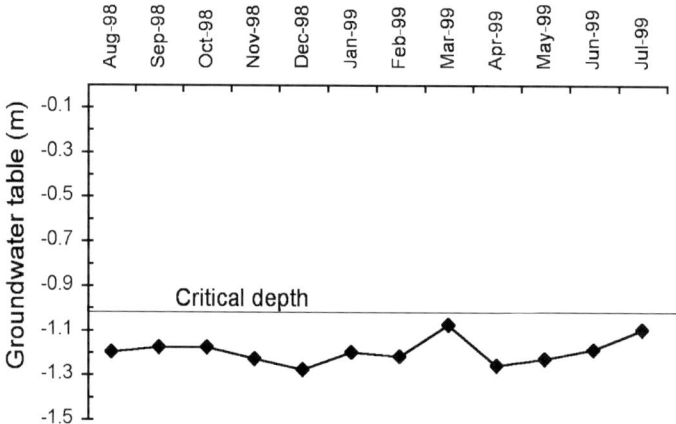

Fig. 3.12. Fluctuation of groundwater depth with respect to its critical value; Nilo Coelho, Brazil (Bastiaanssen *et al.*, 2001).

Table 3.8. Minimum group of recommended pollutants to be monitored.

Type of pollutant	To be measured
Soil salinity	The electrical conductivity (EC) of the soil
Organic matter	The total dissolved organic matter (vol%), floating matter (vol%), colour and smell
Biological matter	Biochemical oxygen demand (m/l) and the chemical oxygen demand (m/l)
Chemicals	We recommend the measurement of at least the concentration of nitrates (NO_3^- in meq/l) and of phosphorus (P in meq/l)

In this section, we only recommend monitoring a group of pollutants whose concentration can be determined at low cost per (laboratory) measurement. We tentatively assume that if none of these parameters have a value approaching critical levels, that other pollutants (e.g. heavy metals, pesticides, etc.) will not cause a problem. This assumption, however, should be checked for the month during which this pollution is anticipated to be highest. The recommended group is shown in Table 3.8.

Sustainability of irrigable area

The intensity with which the irrigated area is cropped traditionally is a function of the number of crops per year grown on an irrigated area. For cropping patterns of various crops with widely different lengths of growing period, and for orchards, however, this cropping intensity is not well defined. To quantify the 'occupancy' of the irrigable area by a crop it is recommended to use the ratio:

$$\text{Cropped area ratio} = \frac{\text{Average cropped area}}{\text{Initial total irrigable area}}$$

The cropped area is the weighted average during the considered period (usually 1 month, see Fig. 3.13). The initial area refers to the total irrigable area during the design of the system or following the latest rehabilitation. If the area ratio is averaged over 1 year, it quantifies the rate at which the irrigable area is occupied by crops. This average area ratio is automatically calculated by CRIWAR (Bos *et al.*, 1996).

Within the irrigated area, several negative impacts (waterlogging, salinity and water shortage due to competitive use) cause a reduction of the (actually) irrigated area. A further reduction of the cropped area is related with population growth and urbanization, road construction, etc. Parameters of physical sustainability (of the irrigated area) that can be affected by irrigation managers relate primarily to over- or under-supply of irrigation water, leading to waterlogging or salinity. The cumulative effect of the above (negative) impacts on the cropped area ratio can be quantified by plotting annual values of this ratio. If the annual average

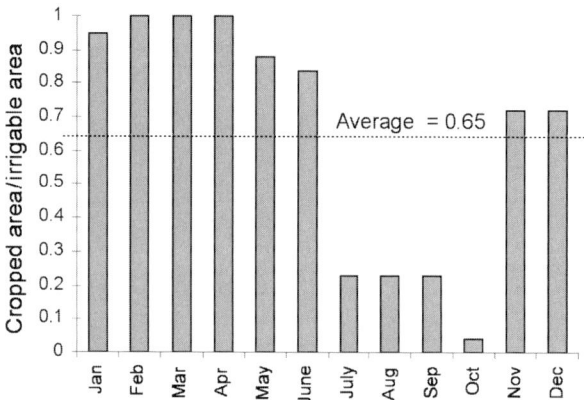

Fig. 3.13. Annual variation and average for the cropped area ratio (Bos et al., 1996).

cropped area ratio is mapped for each tertiary unit, the area with relatively low land occupancy is visualized.

Economics

Each of the primary participants in the irrigation sector, i.e. planners and policy makers, agency personnel and farmers, has a different perspective on what is meant by economic performance. Each, therefore, requires a separate set of indicators that reflects these different objectives. The system manager is most likely to be concerned with the financial resources available at system level and the source of those funds. Policy makers are more concerned with overall returns on resource use from agriculture, and less concerned about the overall profitability of the irrigation institution that created the system (unless it is owned by a private firm in which they are shareholders). Farmers are interested in the returns to their farming enterprise, and less concerned about overall returns to the resource base.

Water productivity

Within many irrigated areas, water is an increasingly scarce resource. Hence, it is logical to assess the productivity of irrigation in terms of this scarce resource (for detailed discussions, see Kijne et al., 2003). Such an assessment can be made from a variety of viewpoints. The most common are: the productivity in terms of actual evapotranspiration and in terms of the volume of supplied irrigation water. The water productivity then is defined as (Molden et al., 1998):

$$\text{Water productivity } (ET) = \frac{\text{Yield of harvested crop}}{ET_{\text{actual}}}$$

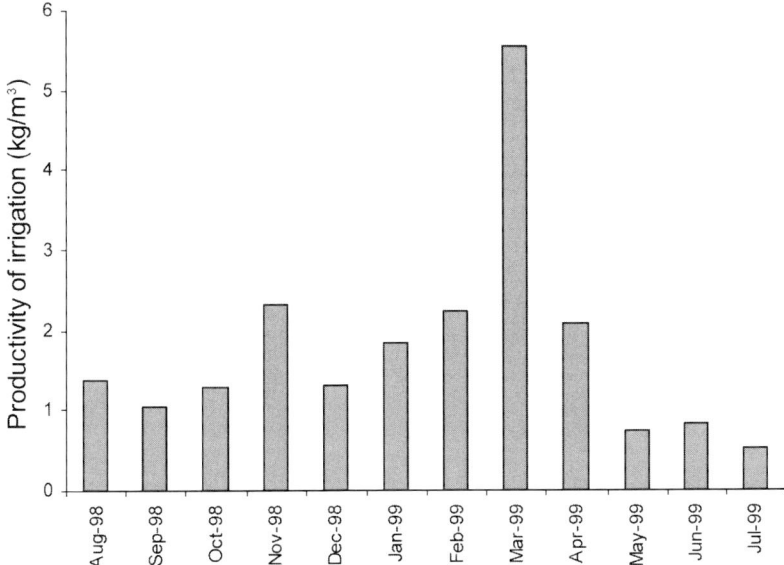

Fig. 3.14. Variation of productivity in terms of kg/m³ irrigation water supplied to the Nilo Coelho scheme, Brazil (Bastiaanssen *et al.*, 2001).

and

$$\text{Water productivity}(m^3) = \frac{\text{Yield of harvested crop}}{\text{Volume of supplied irrigation water}}$$

The yield of the harvested crop equals the unit yield (kg/ha) times the considered area (ha). If viewed from the farmer's perspective, the volume of supplied water is measured either at the farm inlet or at the head of the field, depending on his views. Because the values of ET_{actual} and the volume of (needed) irrigation water are heavily influenced by local climate, they are suited to tracking performance over time as in Fig. 3.15.

Productivity of water can be expressed in terms of monetary value per unit of water. Gross value of production is the yield multiplied by the price of output, while the net value includes costs. This is useful when an irrigation system has multiple crops, especially grain and non-grain, like maize, potatoes and fruits. Increases in economic water productivity may indicate a shift towards higher valued crops or an increase in yields. Figure 3.15 shows the spatial variability of water productivity within the Gediz basin.

Land productivity

Independently of the economic viability of a particular investment, or the viability of the agencies supplying water and other inputs, farmers

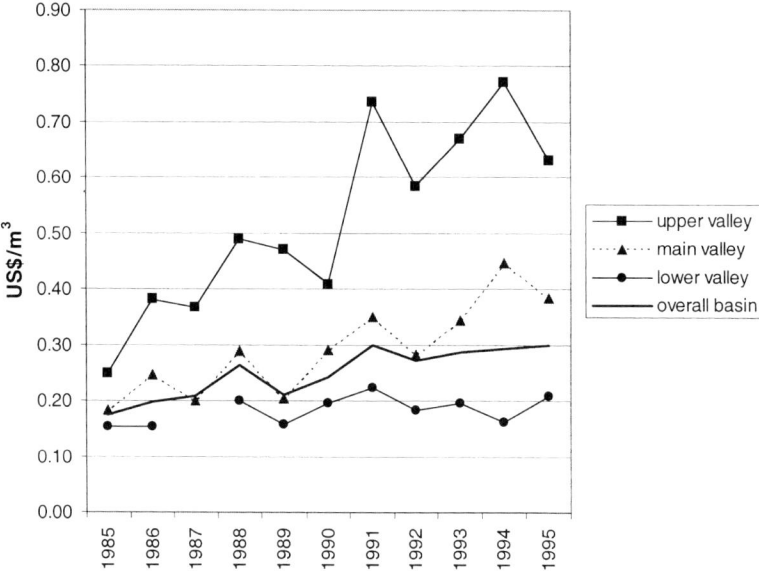

Fig. 3.15. Water productivity per unit of water supplied in different locations of the Gediz Basin, Turkey.

must primarily be concerned with the profitability of their actions at the level of their individual farm. It is quite possible for sector or system level economic analyses to show negative returns, largely through the high cost of capital, and yet find farmers in those systems consistently making profits. This profit is largely determined by crop yield and the farm-gate price of the irrigated crop. To assess crop yield, it should be related to the intended crop yield. This intended yield varies with the crop variety, water application, soil fertility, farm management, etc. The crop yield ratio is:

$$\text{Crop yield ratio} = \frac{\text{Actual crop yield}}{\text{Intended crop yield}}$$

The actual crop yield also can be plotted in its own right against time (as in Fig. 3.2). The most common method to plot data, however, is as a function of space (Fig. 3.16). To assess performance, however, it always must be related to the intended yield.

Financial viability of irrigation systems

One set of indicators is concerned with efforts to raise revenues from water users to help support management, operation and maintenance (MO&M) costs, and often some or all of the capital costs of individual irrigation systems. The first of these indicators describes the overall financial viability of the system:

Fig. 3.16. Satellite-based view showing variations of mango yield with respect to the target yield for the Nilo Coelho project, 1999 (Bastiaanssen et al., 2001).

$$\text{MO\&M funding ratio} = \frac{\text{Actual annual income}}{\text{Budget for sustainable MO\&M}}$$

The total MO&M requirements should be based on a detailed budget which is approved through a good budgeting system. If such a system is not in place, a budget can be based on the estimated MO&M expenditure per hectare. The indicator is admittedly subjective because 'requirements' greatly depend on the number of persons employed by the agency per unit irrigable area (for ranges, see Bos and Nugteren, 1974). However, it gives an indication of the extent to which the agency is expected to be self-financing. The above income of the agency (users' association, irrigation district, irrigation department, etc.) may have different sources of income, e.g. subsidies from central government, water charges, sale of trees along canals, hydraulic energy, etc.

O&M fraction

To quantify the effectiveness of the irrigation agency with respect to the actual delivery of water (system operation) and the maintenance of the canals (or pipelines) and related structures, the O&M fraction is used.

$$\text{O\&M fraction} = \frac{\text{Cost of operation} + \text{maintenance}}{\text{Total budget for sustainable MO\&M}}$$

This indicator deals with the salaries involved with the actual operation (gatemen, etc.) plus maintenance costs and minor investments in the system (replacement of canal or pipe sections and of damaged structures). To quantify the O&M fraction, we need the annual budget as *proposed* by the irrigation authority (for its total O&M) and from the WUA of the selected command area (for its O&M), the budgets as *approved* (allocation per item) and the *actually* realized income over the related year. Table 3.9 gives an example for the Tunuyan scheme, Mendoza, Argentina.

Fee collection ratio

In many irrigated areas, water charges (irrigation fees) are collected from farmers. The fraction of the annual fees (charges) due to be paid to the WUA and/or the irrigation district is an important indicator for level of acceptance of irrigation water delivery as a (public) service to the customers (farmers). The indicator is defined as:

$$\text{Fee collection ratio} = \frac{\text{Irrigation fees collected}}{\text{Irrigation fees due}}$$

Figure 3.17 shows the fee collection ratio for the Nilo Coelho project (15,200 ha) in Brazil. Up to 1994, the fee collection ratio was too low to pay for all required maintenance. With the introduction in 1995 of a strict *no pay* $\Rightarrow$ *no water* policy, the ratio increased to 1.05. Because farmers were paying their arrears, the indicator value is greater than 1.

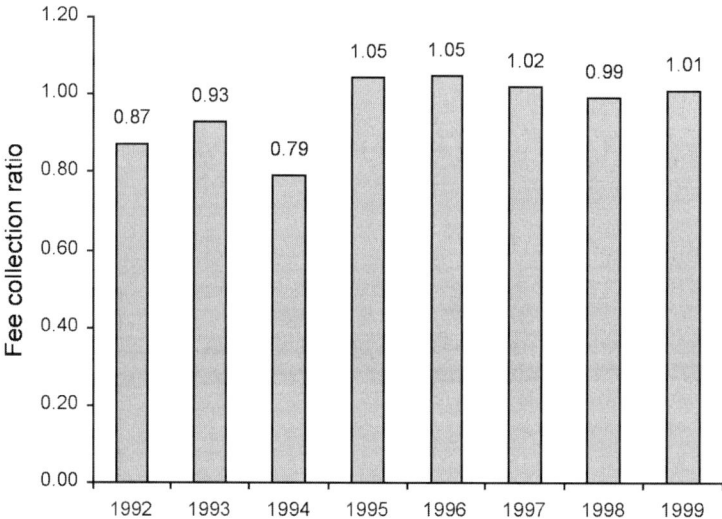

Fig. 3.17. Fee collection ratio for the Nilo Coelho project, Brazil.

Table 3.9. Average weight (%) of items in the UA's budgets in Tunuyan, Argentina (Marre et al., 1998).

Size of UA (ha)	Average O&M fraction in % of budget	O&M items				Management	
		Salaries of O&M personnel	Canal cleaning and maintenance	Minor works	Per diem and transport	Administrative cost	Other budget items
<1,000	65	38	16	11	12	17	6
1,001–3,000	58	33	18	7	20	16	6
3,001–6,000	59	31	24	5	9	11	20
6,001–9,000	61	41	17	3	10	10	19
9,001–12,000	56	42	12	2	13	9	22
>12,000	80	50	20	10	10	2	8
Average ⇒	63	39	18	6	12	11	14

Relative water cost

From the perspective of the farmer, the relative cost of irrigation water application plus the cost of drainage can also quantify the economics of irrigation. The relative water cost equals:

$$\text{Relative water cost} = \frac{\text{Total cost of irrigation water}}{\text{Total production cost of major crop}}$$

The total production cost includes cost of water (including fees, energy for pumping), seeds, fertilizer, pesticides, labour, etc. For surface irrigation, this ratio often ranges between 0.03 and 0.04; if pumped groundwater is used, the ratio may become as high as 0.10. If the ratio becomes higher, farmers may abandon irrigation.

Price ratio

At the end of the irrigation season the farmer needs a 'reasonable' farm gate price for his crop. In this context 'reasonable' is compared with the price of the same crop at the nearest market. The price ratio, which is recommended to quantify this key parameter, is defined as:

$$\text{Price ratio} = \frac{\text{Farm gate price of crop}}{\text{Nearest market price of crop}}$$

Low values of this ratio occur with inadequate distribution and marketing systems and if the distance to the nearest market is long. A low price ratio is a common reason for the farmer to change crop or stop irrigation entirely.

Emerging indicators from remote sensing

The opportunity to measure data through satellite remote sensing became feasible with the cost reduction of images and the advances in software and computers. This combination of developments facilitates studying crop growing conditions at scales ranging from individual fields to scheme or river basin level. Public domain Internet satellite data can be used to calculate actual and potential crop evapotranspiration, soil moisture and biomass growth. Satellite-interpreted raster maps can be merged with vector maps of the irrigation water delivery system and (monthly) values of performance indicators for the various irrigation units (lateral or tertiary) can be presented through standard GIS. The accuracy with which data can be measured compares well with traditional measurements (Chapter 6).

Crop water deficit

Crop water deficit over a period is defined as the difference between the potential and actual evapotranspiration of the cropping pattern within an area as defined by the water manager. A common period is 1 month. Thus:

Crop water deficit = $ET_p - ET_a$ (in mm/month)

Figure 3.18 shows monthly values ranging from 0 to 90 mm/month for the lateral units of the Nilo Coelho project, Brazil.

If an average crop water deficit of 1 mm/day is accepted, i.e. 30 mm/month, then only a few of the lateral units are in the proper range. The availability of data for each pixel allows the computation of the average and standard deviation of the indicator. Also, the percentage of pixels outside the acceptable range of the performance indicator can be calculated. Remote sensing data, thus, are suitable to obtain spatio-temporal information on irrigation performance.

Relative evapotranspiration

To evaluate the adequacy of irrigation water delivery to a selected command area as a function of time, the dimensionless ratio of actual over potential evapotranspiration gives valuable information to the water manager. The ratio is defined as:

$$\text{Relative evapotranspiration} = \frac{ET_{actual}}{ET_{potential}}$$

Figure 3.19 shows average monthly values of the ratio for the Nilo Coelho project (12,849 ha). The indicator is relatively stable near the lower side of the allowable range.

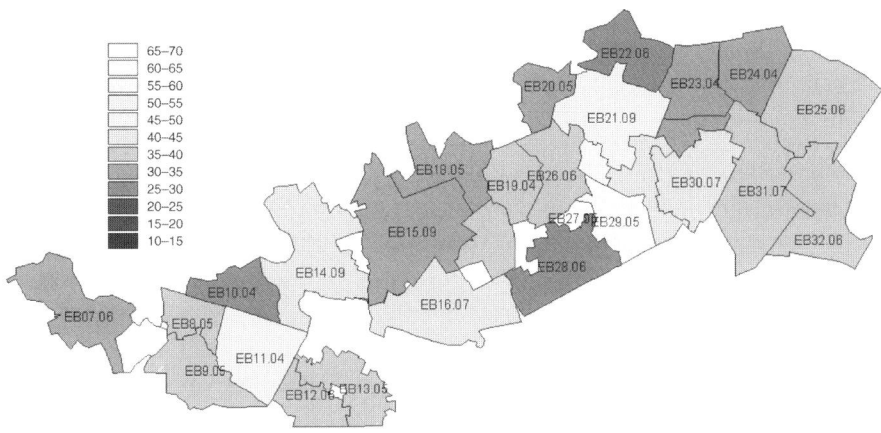

Fig. 3.18. Spatial distribution of the crop water deficit (in mm/month for January 1999) for the lateral unit of the Nilo Coelho scheme (Bastiaanssen *et al.*, 2001).

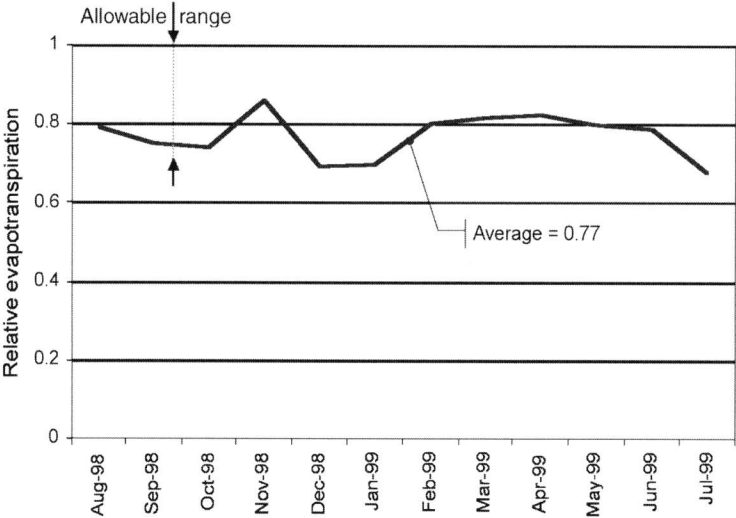

Fig. 3.19. Temporal variation of the relative evapotranspiration with respect to the allowable range for the Nilo Coelho project.

Relative soil wetness

The relative soil wetness is a measure for the ease with which the (irrigated) crop can take water from the root zone. It is defined as:

$$\text{Relative soil wetness} = \frac{\theta_{actual}}{\theta_{FC}}$$

θ_{actual} = measured (actual) volumetric soil water content in the root zone (cm³/cm³); θ_{FC} = volumetric soil water content at field capacity (cm³/cm³).

Figure 3.20 shows that the interannual fluctuations of the relative soil wetness for Nilo Coelho are small, and that the relative soil wetness remains above 1. This implies that the soil moisture remains at, or above, field capacity. The irrigation methods (drip and micro sprinkler) keep water in the root zone at a too 'wet level'. From October to April, less irrigation water can be applied and still allow the crop to take water from the root zone without a potential yield reduction.

Biomass yield over water supply

The biomass yield over irrigation water supply ratio is a surrogate of the productivity of water. It relates the crop growth expressed as above-ground dry biomass growth (kg/ha per month) with the volume of irrigation water supplied to the irrigated area (m³/month). The ratio thus is:

$$\text{Biomass yield over irrigation supply} = \frac{\text{Bio}}{V_C}$$

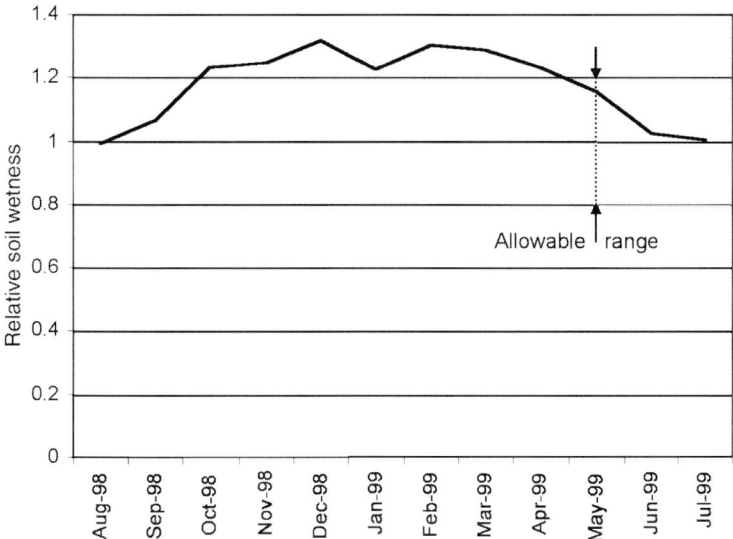

Fig. 3.20. Variation of the relative soil wetness during the irrigation season.

As with crop yield, the spatial variation of biomass production gives valuable management information on water use. Figure 3.21 illustrates the wide variation. A similar chart can be made for each month.

If the average harvest index (harvested crop over biomass production) for a crop is known, the above ratio can be transferred into productivity data.

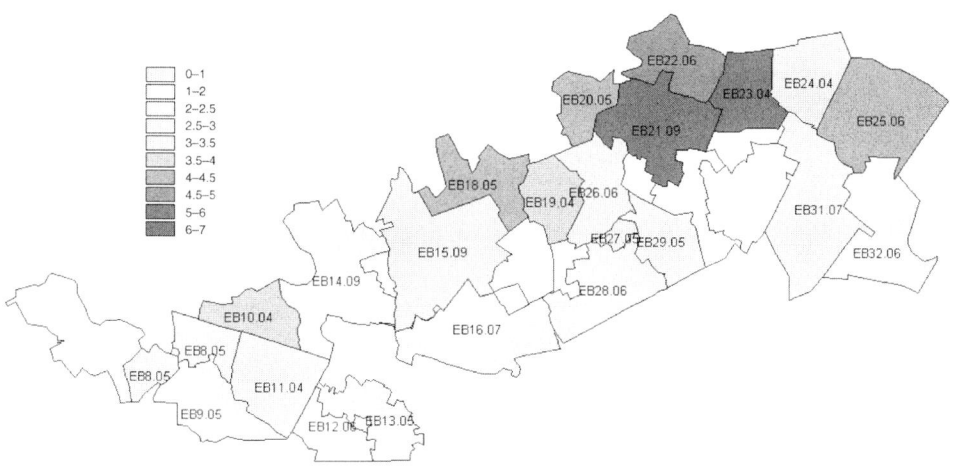

Fig. 3.21. Biomass yield over irrigation supply for February 1998 (kg/m^3) in the Nilo Coelho scheme (Bastiaanssen *et al.*, 2001).

Grouping of Indicators

As discussed in Chapter 2, the level of detail with which performance should be assessed depends on the purpose of the assessment. Researchers tend to assess performance in full detail. Depending on the disciplines involved, the entire long list of indicators may be used. The cost of collection and handling of all related data, however, is not justified for day-to-day operational management of the system. For this purpose, we recommend reducing the number of indicators as much as possible. The selected shortlist of indicators depends on the local (boundary) conditions and on the reason the assessment is done (Chapter 2). For example, policy makers (at river basin planning level and the ministry of irrigation) commonly consider strategic issues, while system managers tend to concentrate on operational matters unless a significant problem needs to be diagnosed. A subdivision of indicators along these functions is shown in Table 3.10.

Table 3.10. Major function (strategic, operational or diagnostic) of selected performance indicators.

	Policy or strategic	Operational	Diagnostic
Water balance, water service and maintenance			
Overall consumed ratio		✓	✓
Field application ratio, drainage		✓	✓
Depleted fraction	✓		✓
Drainage ratio	✓		✓
Outflow over inflow ratios (as a group)		✓	✓
Delivery performance ratio (actual over intended value in terms of discharge, volume and time)		✓	✓
Dependability of interval between water applications		✓	✓
Canal water level and head–discharge ratios			
Effectivity of infrastructure		✓	✓
Environment			
Groundwater depth	✓		✓
Pollution levels of water	✓		✓
Sustainability of irrigable area	✓		✓
Economics			
Water productivity (in kg/m³ ET_a or kg/m³ irrigation water)	✓		✓
Land productivity (in kg/ha command area or kg/ha cropped area)	✓	✓	✓
MO&M funding ratio	✓	✓	
O&M fraction		✓	✓
Fee collection ratio		✓	
Relative water cost			✓
Price ratio	✓		✓

Table 3.10. *Continued.*

	Policy or strategic	Operational	Diagnostic
Emerging indicators			
Crop water deficit	✓	✓	✓
Relative evapotranspiration (evaporative fraction)	✓	✓	✓
Relative soil wetness	✓	✓	✓
Biomass production per m³ water supply	✓	✓	✓

Notes

[1] Levine (1982) defined a similar, but inverted indicator termed the relative water supply.

[2] Consumed refers to crop evapotranspiration, while depleted refers to a use that renders it unavailable for further use within the system or downstream, either through evapotranspiration, evaporation, severe quality degradation, or flows directed by irrigation to sinks.

References

Bastiaanssen, W.G.M. and Bos, M.G. (1999) Irrigation performance indicators based on remotely sensed data: a review of literature. *Irrigation and Drainage Systems* 13, 291–311.

Bastiaanssen, W.G.M., Brito, R.A.L., Bos, M.G., Souza, R., Cavalcanti, E.B. and Bakker, M.M. (2001) Low cost satellite data applied to performance monitoring of the Nilo Coelho irrigation scheme, Brazil. *Irrigation and Drainage Systems* 15, 53–79.

Bos, M.G. (1996) *Proceedings of the NATO Advanced Research Workshop on the Interrelationship between Irrigation, Drainage and the Environment in the Aral Sea Basin.* Kluwer, Dordrecht, The Netherlands.

Bos, M.G. (1997) Performance indicators for irrigation and drainage. *Irrigation and Drainage Systems* 11, 119–137.

Bos, M.G. (2001) Why would we use a GIS DataBase and Remote Sensing in irrigation management? In: van Dijk, A. and Bos, M.G. (eds) *GIS and Remote Sensing Techniques in Land and Water Management.* Kluwer, Dordrecht, The Netherlands, pp. 1–8.

Bos, M.G. and Nugteren, J. (1990) *On Irrigation Efficiencies,* 4th edn. ILRI publication 19. International Institute for Land Reclamation and Improvement, Wageningen, The Netherlands.

Bos, M.G., Wolters, W., Drovandi, A. and Morabito, J.A. (1991) The Viejo Retamo secondary canal – performance evaluation case study: Mendoza, Argentina. *Irrigation and Drainage Systems* 5, 77–88.

Bos, M.G., Abdel-Dayem, S. and Abdel-Rahman Attia, F. (1994a) Assessing performance of irrigation and drainage: examples from Egypt. *Proceedings 8th IWRA World Congress on Water Resources*, Cairo, November 1994. Volume 1, T4-S1, pp. 6.1–6.18.

Bos, M.G., Murray-Rust, D.H., Merrey, D.J., Johnson, H.G. and Snellen, W.B. (1994b) Methodologies for assessing performance of irrigation and drainage management. *Irrigation and Drainage Systems* 7, 231–261.

Bos, M.G., Vos, J. and Feddes, R.A. (1996) *CRIWAR 2.0: A Simulation Model on Crop Irrigation Water Requirements.* ILRI publication 46. International Institute for Land Reclamation and Improvement, Wageningen, The Netherlands.

Bos, M.G., Salatino, S.S. and Billoud, C.G. (2001) The water delivery performance within the Chivilcoy Tertiary Unit. *Irrigation and Drainage Systems* 15, 311–325.

Bustos, R.M., Marre, M., Salatino, S., Chambouleyron, J. and Bos, M.G. (1997) Performance of water users associations in the lower Tunuyan area. Submitted for publication to *Irrigation and Drainage Systems.*

Chambouleyron, J. (1994) Determining the optimal size of water users' associations. *Irrigation and Drainage Systems* 8, 189–199.

Clemmens, A.J. and Bos, M.G. (1990) Statistical methods for irrigation system water delivery performance evaluation. *Irrigation and Drainage Systems* 4, 345–365.

ICID (1978) Standards for the calculation of irrigation efficiencies. *ICID Bulletin* 27, 91–101.

International Irrigation Management Institute (IIMI) (1989) Efficient irrigation management and system turnover. Final Report. Volume 2. ADB Technical Assistance TA 937-INO, Indonesia.

Irrigation Management Policy Support Activity (IMPSA) (1991) *Modernizing the Irrigated Agriculture Sector: Transformations at the Macro-institutional Level.* Policy paper no. 4. IMPSA Secretariat, Colombo, Sri Lanka.

Jurriens, M., Zerihun, D., Boonstra, J. and Feyen, J. (2001) *SURDEV: Surface Irrigation Software.* International Institute for Land Reclamation and Improvement, Wageningen, The Netherlands.

Kijne, J., Barker, R. and Molden, D. (eds) (2003) *Water Productivity in Agriculture: Limits and Opportunities for Improvement.* CAB International, Wallingford, UK.

Marre, M., Bustos, R., Chambouleyron, J. and Bos, M.G. (1997) Irrigation water rates in Mendoza's decentralized irrigation administration. In: van Hofwegen, P.J.M. and Schultz, E. (eds) *Financial Aspects of Water Management.* Balkema Publishers, Rotterdam, pp. 25–40. Revised edition: *Irrigation and Drainage Systems* 12, 67–83.

Merrey, D.J., de Silva, N.G.R. and Sakthivadivel, R. (1992) A participatory approach to building policy consensus: the relevance of the Irrigation Management Policy Support Activity of Sri Lanka for other countries. *IIMI Review* 6, 3–13.

Molden, D.J. (1997) *Accounting for Water Use and Productivity.* SWIM (System Wide Initiative on Water Management) report number 1. International Irrigation Management Institute, Colombo, Sri Lanka.

Molden, D.J. and Gates, T.K. (1990) Performance measures for evaluation of irrigation water delivery systems. *ASCE Journal of Irrigation and Drainage Engineering* 116.

Molden, D.J., Sakthivadivel, R., Perry, C.J., de Fraiture, C. and Kloezen, W.H. (1998) *Indicators for Comparing Performance of Irrigated Agricultural Systems.* Research report 20. International Irrigation Management Institute, Colombo, Sri Lanka.

Morabito, J., Bos, M.G., Vos, S. and Brouwer, R. (1998) The quality of service provided by the Irrigation Department to the users associations, Tunuyan system, Mendoza, Argentina. *Irrigation and Drainage Systems* 12, 49–65.

Murray-Rust, D.H. and Snellen, W.B. (1993) *Irrigation System Performance Assessment and Diagnosis.* (Joint publication of IIMI, ILRI and IHE.) IIMI, Colombo, Sri Lanka.

Renault, D., Hemakumara, M. and Molden, D.J. (2001) Importance of water consumption by perennial vegetation in irrigated areas of the humid tropics: evidence from Sri Lanka. *Agricultural Water Management* 46, 215–230.

Sakthivadivel, R., de Fraiture, C., Molden, D.J., Perry, C. and Kloesen, W. (1999) Indicators of land and water productivity in irrigated agriculture. *Water Resources Development* 15, 161–179.

Small, L. (1992) *Evaluating Irrigation System Performance with Measures of Irrigation Efficiencies.* ODI Irrigation Management Network Paper No. 22. Overseas Development Institute, London.

Small, L.E. and Svendsen, M. (1990) A framework for assessing irrigation performance. *Irrigation and Drainage Systems* 4, 283–312. Revised edition as: Working Paper on Irrigation Performance 1. International Food Policy Research Institute, Washington, DC.

Smit, M. (1989) *CROPWAT: Program to Calculate Irrigation Requirements and Generate Irrigation Schedules.* Irrigation and Drainage Paper 46. FAO, Rome.

Till, M.R. and Bos, M.G. (1985) The influence of uniformity and leaching on the field application efficiency. *ICID Bulletin* 34.

Wolters, W. (1992) *Influences on the Efficiency of Irrigation Water Use.* ILRI publication no. 51. International Institute for Land Reclamation and Improvement, Wageningen, The Netherlands.

Wolters, W. and Bos, M.G. (1990) *Irrigation Performance Assessment and Irrigation Efficiency.* 1989 Annual Report. International Institute for Land Reclamation and Improvement, Wageningen, The Netherlands, pp. 25–37.

4 Operational and Strategic Performance Assessment

Introduction

This chapter looks at performance assessment in the context of the day-to-day functioning of irrigation and drainage systems, and shows how performance assessment can be integrated into the management processes of irrigation and drainage systems.

An explanation of service delivery in irrigation and drainage is provided, followed by a brief discussion of the impact that different formulations of the physical infrastructure and management structure can have on service delivery. Approaches are then formulated for operational and strategic performance assessment of service delivery in these different contexts.

A key focus of the chapter is on performance-oriented management, the basic components of which are:

- The specification of the services and the level of service provision by the irrigation service provider to the water users.
- Agreement between the water users and the irrigation service provider on the rights and responsibilities of the water users, particularly in relation to payment for services received.
- Procedures for monitoring the services provided and responsibilities fulfilled.
- Procedures for evaluating the services provided and the responsibilities fulfilled.

The irrigation and drainage service provider is responsible for the abstraction, conveyance and delivery of irrigation water to the water users, and removal of drainage water. The specification of the level of service to be provided varies, but will generally relate to the reliability, adequacy and timeliness of water delivery and removal. The agreement between the water users and the service provider can be explicitly or implicitly stated – explicitly through a signed contract or legal instrument (laws, statutes or bylaws), or implicitly through convention or historical precedence. As well as the service specification, such agreements generally cover issues of payment for the service provided, and the responsibility of the user to protect and not misuse the irrigation and drainage infrastructure.

In many countries the rules and regulations, and the roles, duties and responsibilities of the various parties, have been set out in the national laws related to irrigation and drainage. Such approaches, which are generalized for a country or region, are being supplemented with service agreements between the service provider and the water users for specific systems, providing a more responsive and accountable relationship between these two parties.

Service Delivery

Understanding service delivery

Two primary functions of the management of irrigation and drainage systems are the supply of irrigation water and the removal of excess water to or from specific locations at specific times. The level to which these functions are to be provided has to be specified in quantitative operational service standards. These standards serve to guide the management activity, and to provide a base against which the performance of the service can be assessed.

The *level of service provision* in irrigation and drainage is defined by Malano and Hofwegen (1999) as:

> A set of operational standards set by the irrigation and drainage organization in consultation with irrigators and the government and other affected parties to manage an irrigation and drainage system.

In principle, the formal specification of the level of service for an irrigation and drainage system emerges from a consultative process between the irrigation and drainage service provider and the water users. In some systems the level of service is clear and explicitly stated, in others it is not. With the greater participation of water users in the management processes, the level of service provision is now being more explicitly formulated for many systems.

The principal elements of service provision (Fig. 4.1) are:

- The provision of the service.
- Payment for the service received.
- The service agreement.

The service agreement includes the *specifications* and *conditions* that detail what service will be provided and what fee the user agrees to pay.

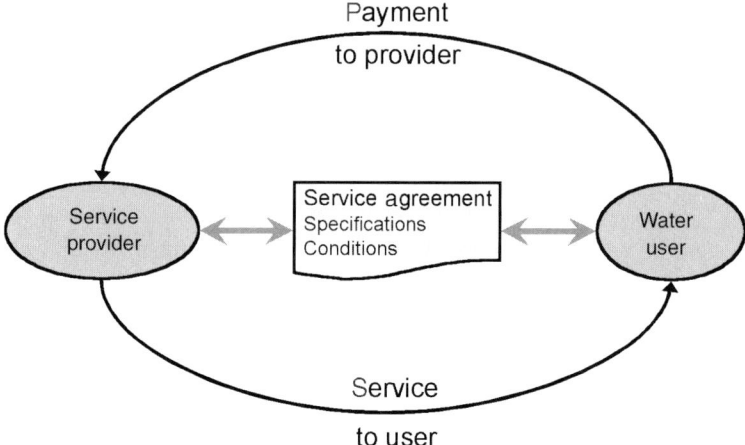

Fig. 4.1. Interaction of core elements of service delivery (Huppert and Urban, 1998).

The specification sets out the services that will be provided, and the standard to which those services will be provided (for example, the provision of irrigation water within 24 h of receipt of the water user's request, or drainage of land within 24 h of heavy rainfall). The conditions stipulate the terms under which the service will be provided (for example, that fees will be paid for irrigation water received or drainage water removed).

The service agreement generally takes the form of an agreement between two parties. In the case of water users' associations (WUA), the service delivery agreement between the WUA and the water users is often specified within the statutes and bylaws. Through this process the water user is aware of their rights (in terms of access to, and receipt of, water), and responsibilities (payment or contribution in kind – for example, for maintenance), and can hold the service provider responsible for meeting the agreed service standards. Through this process the delivery of irrigation water and/or the removal of drainage water becomes more transparent and accountable. Performance assessment is a key component of the process in holding each party accountable.

The institutional aspects of service delivery form an often unseen but crucial part of the relationships outlined in Fig. 4.1. Too often the principal focus in performance assessment is on the technical aspects (measurement of frequency, rate and duration of water supply), yet the institutional aspects, such as the legal framework, management decision making or social attitudes, can fundamentally undermine the proper functioning of service provision.

Technically strong systems will often fail to deliver if the institutional arrangements are inadequate. In contrast, systems with low levels of technology are made to work well in institutionally strong environments (for example, the *subak* system in Bali or the hill irrigation systems in Nepal).

Performance assessment must therefore take account of qualitative, as well as quantitative aspects of system management. As discussed in Chapter 2, such considerations must be taken into account when setting the boundaries of the performance assessment programme.

Formulating specifications for service delivery

Irrigation and drainage schemes have been developed with many different objectives. Some schemes have been designed with the primary objective of flood protection, with drainage and irrigation a secondary objective. Others have been designed for 'protective irrigation', providing a minimum level of irrigation supply to protect against drought. Yet others have been developed as commercial enterprises.

Service specifications describe how services will be delivered to meet objectives. For example, water may be delivered at fixed intervals. Alternatively, water may be delivered as per user demands. Each type of service has various advantages, and associated costs, and may be adapted to local situations. For example, delivery of water on a rotational basis may be satisfactory for rice-growing areas, but may not be adequate for vegetable-producing areas, where more flexible water deliveries are required. The cost of the rotational service is likely to be much cheaper than the on-demand service, which requires more structures for water control and measurement, and generally more intense management.

Replogle and Merriam (1980) have outlined a useful categorization of irrigation service delivery schedules based on the three variables of *frequency*, *rate* and *duration* (Table 4.1). These three variables are governed by the conveyance systems and control structures. For simple run-of-the-river systems with limited control systems, the frequency that the water user receives water is fixed (constant flow), the rate is governed by the discharge in the river and the duration is also fixed (constant flow). For an on-demand system, the frequency that the water user receives water can be varied, as can the rate and the duration. The only possible limitation in this case might be on the design capacity of the canal or pipeline supplying the water.

The full range of irrigation schedules can be defined by these three variables, ranging from on demand, where the frequency, rate and duration of flow are not limited, through to a constant amount – constant frequency schedule, where the frequency, rate and duration are all fixed. An on-demand or limited rate demand schedule is often provided by automated systems, while arranged or limited rate arranged schedules are provided by irrigation systems with variable control gates and measuring structures, and constant amount – constant frequency schedules

Table 4.1. Classification of irrigation schedules (Replogle and Merriam, 1980).

Schedule name	Frequency	Rate	Duration
On demand	Unlimited	Unlimited	Unlimited
Limited rate, demand	Unlimited	Limited	Unlimited
Arranged	Arranged	Unlimited	Unlimited
Limited rate, arranged	Arranged	Limited	Unlimited
Restricted-arranged	Arranged	Constant	Constant
Fixed duration, restricted-arranged	Arranged	Constant	Fixed by policy
Varied amount, constant frequency (modified amount rotation)	Fixed	Varied as fixed	Fixed
Constant amount, varied frequency (modified frequency rotation)	Varied as fixed	Fixed	Fixed
Constant amount, constant frequency (full supply-orientated rotation)	Fixed	Fixed	Fixed

Unlimited: Unlimited and controlled by the user. Limited: Maximum flow rate limited by physical size of offtake capacity but causing only moderate to negligible problems in on-farm operation. The applied rate is controlled by the user and may be varied as desired. Arranged: Day or days of water availability are arranged between the irrigation service provider and the user. Constant: The condition of the rate or duration remains constant as arranged during the specific irrigation turn. Fixed: The condition is predetermined by the irrigation service provider or the system design.

are provided by irrigation systems with limited control gates and/or fixed proportional division structures.

Thus, the technical configuration of the irrigation and drainage system strongly influences the level of service that can be achieved, and therefore the nature and boundaries of any performance assessment programme.

Some service specifications are relatively standard, and can be applied to all irrigation and drainage systems (such as water quality standards, Table 4.2). Other service specifications are more site-specific and have to be formulated based on the particular circumstances for individual schemes. A comparison of different service specifications for four irrigation schemes is presented in Table 4.3, showing some of the key elements considered when setting service specifications, and incorporating the irrigation delivery specifications outlined in Table 4.1.

Finally, it is important to note that objectives change over time in response to changing needs of users or society. For example, users may want the capability to grow higher valued crops, requiring that their system and associated processes are modernized. Or society may demand that irrigation uses less water, forcing a new situation on management. Changing of objectives is part of strategic performance assessment, and is dealt with later in the chapter.

Table 4.2. Examples of water quality standards for surface waters (Malano and van Hofwegen, 1999).

Parameter		Maximum value
General	Appearance/odour	Water free from visible pollution and smell
	Temperature	<25°C
	O_2	5 mg/l
	pH	6–9
Nutrients	P	0.15 mg/l
	N	2.2 mg/l
	Chlorophyll	100 µg/l
	Ammonia	0.02 mg/l
Salts	Chlorides	200 mg Cl/l
	Fluorides	15 mg F/l
	Bromide	8 mg Br/l
	Sulphate	100 mg SO_4/l

Applying Performance Assessment to Different Types of Irrigation and Drainage Systems

Overview

One of the differences, and difficulties, with irrigation and drainage in comparison with other service delivery systems, such as electricity and potable water supply, is the wide variation in the types of irrigation and drainage systems. The variation is across the board, from the climatic conditions, the type of water source, the water availability, the design of the physical infrastructure, the farming system, the social and institutional context, the market availability, the local and national economy, etc.

As discussed in the previous section, two key factors affecting irrigation and drainage service delivery are the configuration of the physical infrastructure and the management processes, both of which effect *control* over the processes involved.

Figure 4.2 outlines the areas where control needs to be exerted to provide a reliable, adequate and timely irrigation water supply and effective drainage, and the potential benefits of such control. The management of the physical infrastructure leads to the provision of water for irrigation and drainage of excess water; this in turn leads to agricultural crop production and farmer income, some of which can then be used to pay for the service provided. Within the internal processes of the service provider, financial, operation and maintenance control systems are required to support the delivery of the service.

Table 4.3. Summary of level of service specifications for four types of irrigation schemes (Malano and Hofwegen, 1999).

Service specification	Societe du Canal de Provence France	Goulburn-Murray Irrigation District, Australia	Triffa Irrigation Scheme, ORMVA de la Molouya, Morocco	Warabandi schemes, Northern India
Type of organization	Service provider: public corporation, shares owned by local government, banks and Chambers of Agriculture. Water uses: private farms, generally moderate size (10–100 ha). Infrastructure franchised by government to service provider	Service provider: public corporation. Water users: private farms, generally large size (>100 ha). Infrastructure owned by government	Service provider: public corporation. Water users: private farms, generally moderate size (5–50 ha). Infrastructure owned by government	Service provider: government agency. Water users: private farms, generally small size (< 5 ha). Infrastructure owned by government
Operational concept	On demand: unlimited	Limited rate arranged	Restricted arranged	Full supply-oriented rotation
Frequency	Unlimited	Arranged (with 4 days' notice)	Arranged. Number of deliveries related to availability of water	Fixed
Flow rate	Unlimited up to maximum	Constrained (by channel capacity)	Constant flow: 20, 30 or 40 l/s	Fixed
Duration	Unlimited	Unlimited	Fixed by agreement: maximum duration based on crop and flow rate	Fixed
Height of supply (command)	Design canal water levels and pipe pressures	Design water level in channel	Design water level in canal	Design water level in secondary canal (FSL – full supply level)
Operation monitoring	Deliveries monitored through volumetric flow meters. Monthly readings taken	The agency ensures that the planned flow rate is delivered, provided customers adhere to scheduled start and finish times. Flow measured with volumetric flow Dethridge meter	Farmers sign receipt after delivery	Deliveries monitored against published schedule for the season

Operational and Strategic Performance Assessment

Delivery performance	According to the service contract, with different service contracts for different uses. Target: 96% of the time low pressure delivered unless stated otherwise	Target: 86% of orders delivered on day requested	Target: delivery in full in accordance with agreed irrigation schedule	Target: deliveries in accordance with published irrigation schedule
Water charges	Fixed + volumetric charge. Fixed charge based on delivery rate. Volume charge based on volume delivered. Gross average: US$0.10/m^3, which includes full cost recovery and asset renewal	Volumetric: US$0.021/m^3. Full cost recovery, including asset renewal	Volumetric: price varies for gravity, lift and pressurized water from US$0.020/m^3 to US$0.040/m^3. Government subsidy to cover cost recovery deficit	Based on crop type and area irrigated. Charge not related to O&M cost; does not cover full O&M cost, cost recovery or asset renewal
Points of supply	One point of delivery per point per contract holder	One point of supply per property	One supply point per group of farmers; farmers rotate supply	One point of supply (head of watercourse) for each group of farmers
Water ordering	On demand, so no ordering necessary	Telephone ordering system – 4 days' notice required	Agency announces an irrigation cycle, farmers can request time and duration of delivery. Schedules are then drawn up and agreed on by all parties	None, supply-oriented. Irrigation schedule (by rotation) is drawn up by the agency at the start of the season and published
Supply restrictions	In case of a water shortage a system of water orders is introduced and allocations are made in proportion to these orders	If demand exceeds available supply, water is allocated equitably to all customers	Prior to season any restrictions for crop types are announced. During the season equitable distribution between permitted crops	In case of a water shortage, rotation schedule is adjusted to reduce (equally) the supply to each secondary canal
Water rights	According to the contracts	Transferable, either temporarily or permanently	Attached to landownership	Attached to landownership

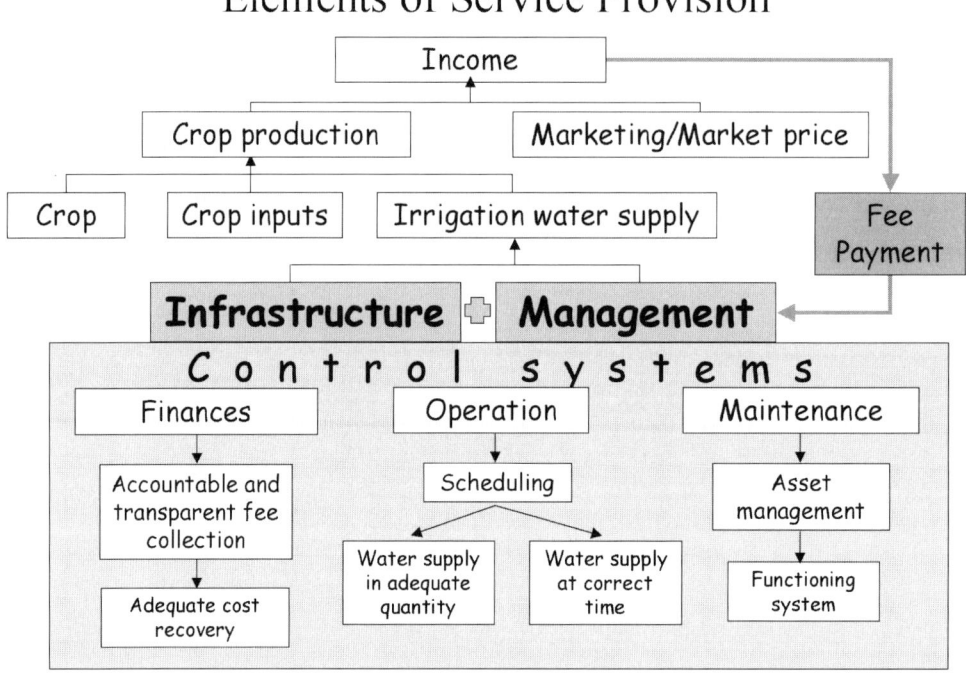

Fig. 4.2. Components of level of service provision to water users.

The level of physical control and measurement built into the irrigation and drainage system design has a fundamental impact on the level and type of operational performance assessment that is: (i) required and (ii) possible. In general, the need for operational performance monitoring increases as the level of control and measurement increases.

Physical characteristics

Figure 4.3 shows some of the components of different types of irrigation and drainage systems. For gravity flow systems, water is diverted from the river into the canal network. Control structures along the way divert, head up and measure the water en route to the farmer's field. Alternatively, water can be pumped from the river and distributed by open channels or closed pipe systems. At field level different methods, ranging from furrow to drip, are used to apply the water to the crop. Surface drains are required to remove rainfall and excess irrigation water, buried drainage systems may be required where the water table rises towards the soil surface. For groundwater systems water is raised by pump and may then be distributed by open channels to the fields, or closed pipe systems.

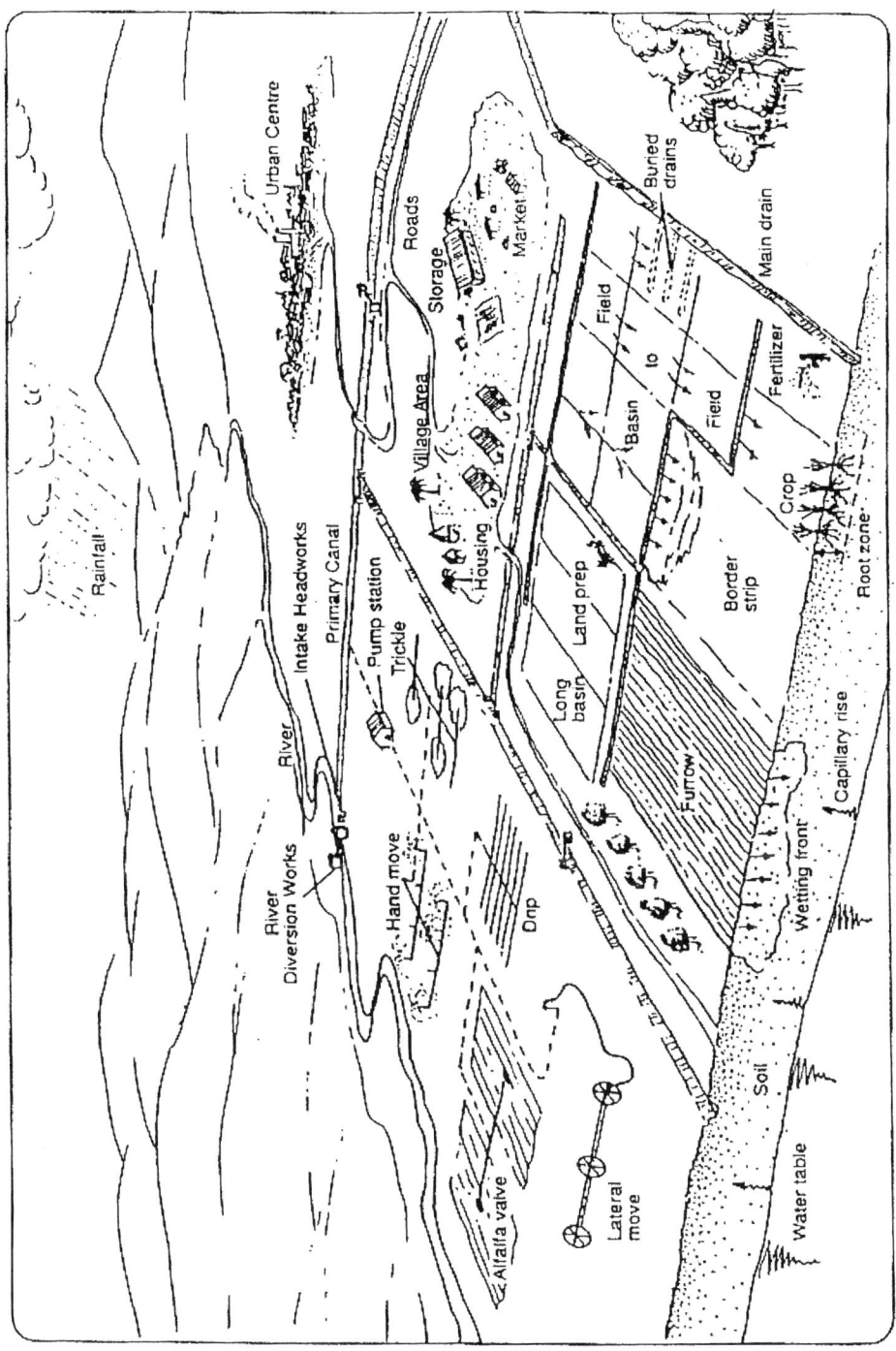

Fig. 4.3. Components of an irrigated farming system.

In the service delivery context the degree of control available at the following points is important:

- Abstraction.
- Conveyance and distribution.
- Application.
- Removal.

The degree of control at the point of abstraction controls the water availability within the system. A system supplied directly from a reservoir will have a different pattern and reliability of water availability to that from a run-of-the-river system. A run-of-the-river system with a river weir and gated headworks will have better control over the water abstraction than a system with a simple diversion channel. Pumped abstraction from groundwater often provides good control of the irrigation water supply.

The type of control (and measurement) structure within the conveyance and distribution system strongly governs the irrigation scheduling that is possible (as discussed earlier), and hence the level of service provision. There is an anomaly here, in that some of the simplest control systems, such as the proportional division weirs used in the hill irrigation systems in Nepal, can provide some of the most reliable levels of service delivery with a fixed frequency, fixed rate and fixed duration supply. Similarly the Warabandi system used in northern India and Pakistan can also provide a reliable level of service delivery based on proportional division of the available water supplies. As one moves to the more sophisticated systems, with cross-regulation structures, gates and measuring structures, the potential for more flexibility exists, and, if managed well, facilitates high levels of production by supplying water either on demand or by arrangement. If managed poorly these systems enable top-end farmers to capture the available water supplies (by manipulating the control structures) at the expense of tail-end farmers. The ultimate in control are automated systems, using techniques such as downstream control based on hydraulic connectivity, or centrally managed networks where control structures are regulated via telemetry or landline communication systems connected to a central computer.

There are a wide variety of application processes, ranging from wild flooding to drip irrigation. The application type will strongly affect the performance at this level, wild flooding, for example, being generally less effective and efficient than drip irrigation for controlled application of the required quantity of water at the crop root zone. Poor control can result in over-application and lead to waterlogging and salinization, at which point buried drainage systems may be required.

Drains are used both to remove excess water and to control groundwater level, and, as with the irrigation channels, require regular maintenance to remain effective.

Measurement of water plays an important part in service delivery; the ability, or inability, to measure water at key points within an irrigation and drainage network governs the management processes and the level of service that can be provided.

Management characteristics

The management type, structure, processes and procedures have a significant impact on service delivery. The design of the physical system sets what is possible, the management processes make it happen.

Different types of management exist, from systems managed entirely from abstraction to application by government organizations, to systems managed entirely by water users' associations or a single private company. Under irrigation management transfer (IMT) programmes, an increasing number of irrigation and drainage systems are being transferred from government agency management to management by water users.

The management structure governs the level of control that can be exerted on the system. With a management structure as might be found on a privately run sugar estate, the general manager has direct control through line management to the field worker applying water to the crops. With a jointly managed irrigation system, where the government agency manages the main canals and the farmers manage within the tertiary units, the government agency only has control of the water to the delivery point at the tertiary intake; the use of water thereafter is under the control of the farmers.

The processes and procedures used by the management to plan, allocate, distribute, monitor and evaluate the irrigation water supply govern how effectively irrigation supply is matched to demand. In some systems, such as the Nepal hill irrigation systems, the processes and procedures are very simple. More sophisticated systems, such as the Warabandi system in northern India and Pakistan, regulate the water distribution within the tertiary unit through predetermined time rosters, though the water is delivered to the tertiary unit (watercourse) on a proportional division basis. As one gets into manually operated gated control systems the need for defined management processes and procedures increases, with decisions needed to be made at regular intervals during the irrigation season to determine irrigation water demands and water allocations at control points. In such systems a fundamental management process is the adjustment of control structures at regular intervals to pass the prescribed discharges.

In manually operated systems the breakdown of the management processes often results in unreliable, inadequate and untimely delivery of irrigation water in relation to the water users' needs. This breakdown can be because of poor management procedures, but can also be due to lack of motivation and incentive for management personnel. Automated irrigation systems are not so reliant on management processes for operation, but do require particular attention being paid to maintenance.

Implementing Operational and Strategic Performance Assessment

Strategic performance assessment

The basic management cycle in an operational context for an irrigation and drainage system is shown in Fig. 4.4. The overall strategic objectives for the system are identified and targets set. These objectives and targets generally apply over a period of several seasons or years, though their relevance may be reviewed on an annual basis. The service agreement is generally negotiated and agreed between the service provider in a similar pattern, though there may be annual or seasonal adjustments to allow for variations in climate, planned cropping, etc.

Performance criteria and indicators can then be formulated which enable the monitoring and evaluation of the achievement of the agreed objectives and targets, as well as the attainment of the conditions of the service specification.

Monitoring and evaluation of scheme performance is carried out during the cropping season or year, and as discussed in Chapter 1 can be of a strategic ('Am I doing the right thing?') or an operational ('Am I doing things right?') nature. Strategic performance assessment is typically done at longer intervals and looks at criteria of productivity, profitability, sustainability and environmental impact. It may also be required in response to changes in the external environment, such as is the case with governments reducing the funding available for supporting irrigated agriculture and transferring responsibility for management, operation and maintenance to water users.

Indicators for strategic performance assessment may differ from those used for operational performance monitoring as they are used to

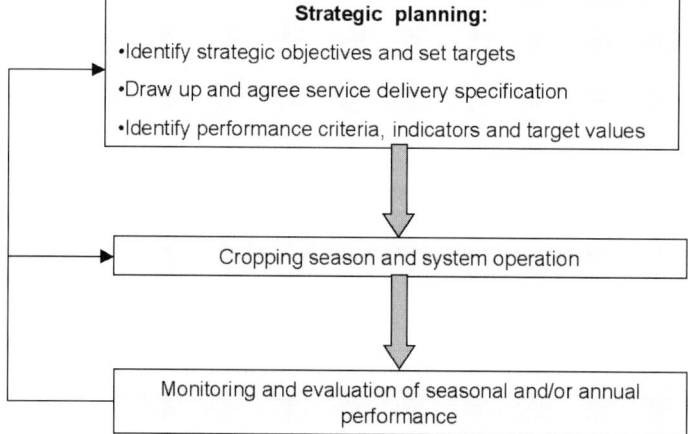

Fig. 4.4. Operational management cycle.

assess changes that may be occurring gradually over time (for example, rise in groundwater levels, salinity or pollution loads).

Operational performance assessment

In order to discuss operational performance assessment it is first necessary to outline some basic operational procedures (Fig. 4.5).

Prior to the commencement of the irrigation season, a pre-season plan is drawn up covering key aspects of the management, operation or maintenance of the system. Depending on the type of irrigation and drainage scheme, this plan covers planned crop areas, estimates of seasonal irrigation water demand and availability, maintenance plans, fee recovery estimates, etc. Budgeting and maintenance work programming are key parts of the planning process. Targets for operational performance assessment are derived from this pre-season plan.

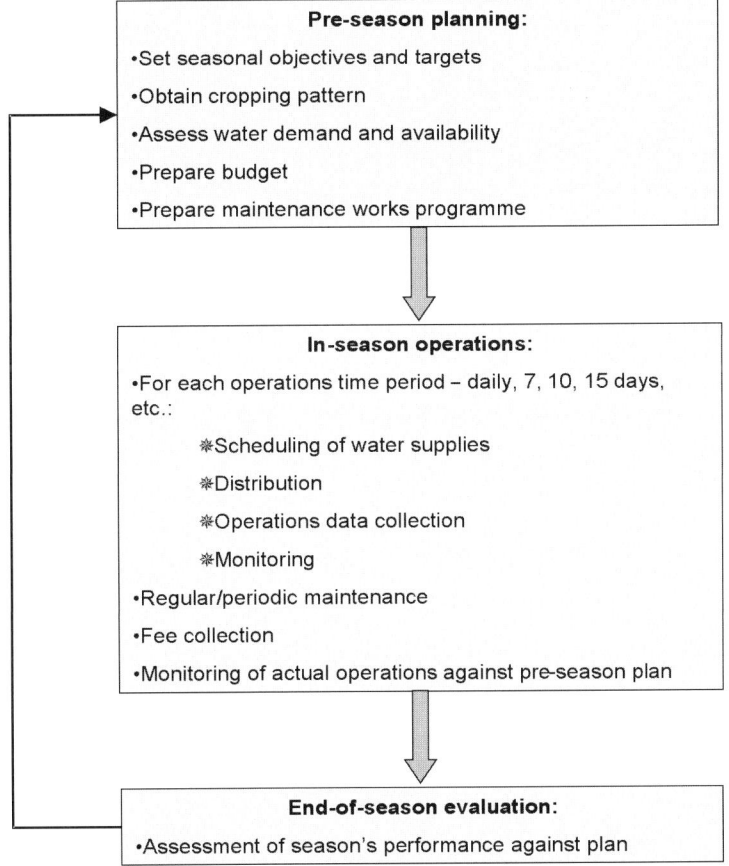

Fig. 4.5. Irrigation management cycle.

The plan is implemented during the season, with further planning being carried out each time period to allocate and schedule irrigation water based on actual irrigation demands and climatic conditions, and to make adjustments to compensate for unplanned events, such as flooding, canal breaches or emergency maintenance. Operational performance assessment carried out during the season supports this planning and adjustment process. The flows in the canal network are regulated in accordance with the implementation schedule and the discharges (and for some schemes, the crop areas) monitored as the season progresses.

The performance of the system in relation to the seasonal plan is monitored during the season, and evaluated at the end of the season. The evaluation measures the performance against the seasonal plan, but may also measure the performance against the strategic objectives.

There is increased demand for transparency and accountability in relation to water management. It is important, therefore, that the findings of the performance monitoring and evaluation process are disseminated to key stakeholders, particularly water users. In schemes that are managed by water users' associations, seasonal performance will be reported at the annual general meeting. For government agency-run systems, seasonal performance results can be published in local newspapers, or displayed in local government offices.

Examples of the different approaches to operational performance assessment based on the type of irrigation and drainage system are outlined in Table 4.4.

Steps in strategic and operational performance assessment

Strategic and operational performance assessment follows the framework outlined in Chapter 2. It contains the following steps:

1. Identification of purpose and extent.
2. Selection of performance assessment criteria, indicators and targets.
3. Data collection.
4. Processing and analysis of data.
5. Reporting results.
6. Acting on results.

The strategic and operational performance assessment procedures are tied into the day-to-day management procedures for the irrigation system. In particular, the data collection, processing and analysis procedures for performance assessment have to be based on the data collected and used for the system management, operation and maintenance. In some cases the data collection, processing and analysis may need to be extended to facilitate better assessment of performance.

1. Identification of purpose and extent

The purpose will be to assess the performance of the irrigation and drainage system in relation to the specifications given in the service agreement (or similar specification of objectives and targets). The boundaries for the irrigation service provider will be the point of abstraction to the point of delivery to the farmer. If a drainage service is provided, the boundaries will include the drainage network and disposal system. The performance assessment is for scheme management, from the perspective of scheme management and farmers (as agreed in the service agreement), is carried out by the scheme personnel and is an operational and accountability type of assessment.

2. Selection of performance assessment criteria, indicators and targets

Performance criteria and indicators are defined based on three considerations:

- Service specification and accountability.
- Strategic objectives.
- Operation and maintenance considerations.

When selecting indicators, consideration is required for how these will be reported, the cost of collecting the information to put in the indicator and the message that the indicator is relaying.

- A key set of indicators will be related to water service delivery. If the specifications call for an arranged schedule, then indicators will be chosen to reflect whether the water request was delivered in the right amount and on time. If the specification calls for constant amount, constant frequency schedule (proportional delivery), some means must be established to ascertain whether flows existed, and whether they are being properly divided. In the first case, the delivery performance ratio, with its average value, and variation over space and time, would be an indicator of delivery performance. In the second case, a proportional dividing structure provides a simple and transparent means of dividing water. As long as water is in the canal, it will be divided. There may not be a need for a formal evaluation of the indicator.
  A key consideration is accountability. There should be a means for both provider and user to ensure that the service is met. The measurement and the indicator should provide for this cross-checking. In the first case, a flow measuring device that both parties can inspect serves the purpose. In the second case, the flow division structure provides a transparent means of assessing whether service has been delivered.
- Strategic performance monitoring is typically done at longer intervals and looks at criteria of productivity, profitability and environmental sustainability. For example, a strategic monitoring programme may

Table 4.4. Linkage between type of irrigation system and operational performance assessment.

System type (as per Table 4.1)	Description	Example location	Control structures	Measuring structures	Cropping	Technology level	Staffing level	Operational planning	Operations data collection	Operational performance assessment
Proportional distribution (Constant amount – constant frequency)	Water distributed in proportion to opening – used in hill irrigation systems in Nepal	Hill irrigation, Nepal	Simple ungated proportional division structures	None	Arrange cropping pattern to match supply pattern	Low	Low	None	None	Monitor structures and ensure no blockages. Volume delivered controlled at design stage by proportional size of opening. The primary objective is equitable distribution of available supplies
	Water distribution on main system in proportion to cultivable command area (CCA). Water allocation within tertiary unit (watercourse) allocated on a time-share basis in proportion to the area of each farmer's plot	Warabandi system, Northern India and Pakistan	Adjustable Proportional Module (APM) at watercourse intake. Simple on/off division boxes in field	Slotted flume on tail of secondary canal (distributary)	Arrange cropping pattern to match average annual water supply pattern	Medium	Low	Medium (to prepare seasonal Warabandi schedule)	Limited (plot and watercourse command areas)	The design requires that the secondary canal flows at design discharge (Full Supply Level, FSL) in order to maintain command over the APM. Canal water levels are monitored at the head of the secondary canal. Frequency and duration of supply to each farmer monitored within the tertiary unit. Rate not monitored
Relative crop area method (Restricted arranged)	Water allocated based on factoring the crop area in relation to the crop's water requirement relative to the base crop. Used in Indonesia, referred to as the Pasten method	East Java, Indonesia	Gated control structures	Required	Varied	High	High, but relatively low skill levels needed for O&M	High	High	Weekly or 10-daily planning of water allocation based on calculated demand. If water short, reduce supply equally to all users. Monitor discharges at primary, secondary and tertiary intakes, compare

Operational and Strategic Performance Assessment

Limited rate, arranged	Water allocated based on calculations of irrigation water demand using standard calculation procedures such as water balance sheets and climatic data	Golbourn-Murray, Australia	Gated control structures	Required	Varied	High	High	High	Regular daily updating of irrigation water demand and planning of water allocation. Water distributed to match demand. The primary objective is to match supply with demand. actual water delivered with plan each week/10 days. Equitable distribution of available water the primary objective, followed by secondary objective of delivering adequate supplies (when water available)
Demand	Water distributed in response to opening of the outlet gates to farms	Aix-en-Provence, France	Automated control structures	Required	Varied	Very high	Low number, but high skill levels	Low	High, but automated	Continuous monitoring of water levels and discharges through automated control systems. Immediate response to irrigation demand. Monitor system to ensure control systems are functioning, and monitor to ensure that total demand can be matched by available supply at water source

investigate changes in groundwater levels, salinity, pollution loads and productivity over a period of several years.
- Other types of indicators may be selected to aid operation and maintenance procedures. For example, measurements of drainage outflow, or condition of structures, will help managers to identify possible causes for the failure to attain specified service levels.

3. Data collection

Maintaining a data collection and monitoring system is necessary to calculate indicators and to provide feedback to users. This is the topic of Chapter 6, and will not be discussed in detail here. For the proportional division system, data collection needs of flow rates may be minimal, whereas in the arranged demand system, more information will be needed about flows as the season progresses. In both cases, financial data on payments and labour contributions are essential.

4. Processing and analysis of data

Data need to be processed and analysed on a regular basis in order to feedback into the management loop. In better resourced systems computers are a standard part of performance management systems, for less well-resourced systems simple processing and monitoring tools, such as operational schematic maps, have a key role to play.

Data processing and analysis is a central feature of irrigation management. In many schemes periodic meetings (weekly, 10-daily, bi-monthly) are held with system managers and staff to: (i) monitor and evaluate performance for the previous time period, and (ii) plan the coming time period's irrigation water allocation and schedule.

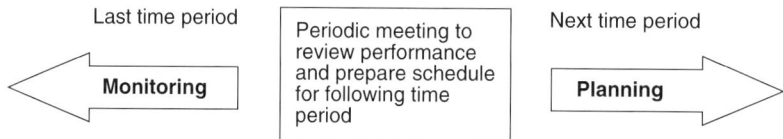

Data are collected on the irrigation demands for the coming time period, and at the end of this time period the supply allocated is compared with the planned allocation. Simple tabulation of the data assists the data processing and analysis. Table 4.5 provides an example where the data sheet is used to calculate the discharge allocations for the coming time period, and then used at the end of the time period to record the actual deliveries and calculate the performance indicators. Figure 4.6 provides a graphical representation of the data, using shading to highlight areas of adequate, over- or under-supply.

Table 4.5. Example of a data processing and analysis form for weekly water allocations.

FORM 04
Division: Region 3
Period: From 22.7.2002 to 29.7.2002

WATER REQUEST, ALLOCATION AND ACTUAL SUPPLY SUMMARY
Canal name: B3 Branch Canal

Note: These last columns are completed at the end of the period

Water Users' Association	Primary/ secondary canal	Command area (ha)	Design canal capacity (l/s)	REQUEST Area irrigated (ha)	REQUEST Discharge (l/s)	REQUEST Duration (h)	PLANNED ALLOCATION Discharge (l/s)	PLANNED ALLOCATION Duration (h)	PLANNED ALLOCATION Handover discharge (l/s)	ACTUAL Discharge (l/s)	ACTUAL Duration (days or h)	MONITORING Delivery performance ratio (actual/ planned)
Col.1	Col.2	Col.3	Col.4	Col.5	Col.6	Col.7	Col.8	Col.9	Col.10	Col.11	Col.12	Col.11/Col.8
	B3	1668	2852	236	1282	24	1282	24	1282	1273	24	0.99
Cane Grove	B3-1	110	132	20	66	24	66	24		64	24	0.97
	B3-2	90	108	18	60	24	60	24		70	24	1.17
	B3-3	80	96	15	50	24	50	24		60	24	1.21
	Sub-total	**280**	–	**53**	**175**	**24**	**175**	**24**	1031	**194**	**24**	**1.11**
Crabwood Creek	B3-4	140	168	17	56	24	56	24		60	24	1.08
	B3-5	167	200	20	66	24	66	24		61	24	0.92
	B3-6	125	150	15	50	24	50	24		62	24	1.25
	B3-7	170	204	20	68	24	68	24		70	24	1.04
	Sub-total	**602**	–	**72**	**239**	**24**	**239**	**24**	689	**253**	**24**	**1.06**
Fellowship	B3-8	102	122	18	60	24	60	24		48	24	0.81
	B3-9	50	60	15	50	24	50	24		53	24	1.07
	B3-10	240	288	29	95	24	95	24		97	24	1.02
	B3-11	65	78	14	46	24	46	24		52	24	1.12
	Sub-total	**457**	–	**76**	**251**	**24**	**251**	**24**	331	**250**	**24**	**1.00**
Golden Grove	B3-12	95	114	18	60	24	60	24		54	24	0.91
	B3-13	54	65	12	40	24	40	24		35	24	0.88
	B3-14	95	114	21	70	24	70	24		55	24	0.79
	B3-15	85	102	19	63	24	63	24		50	24	0.80
	Sub-total	**329**	–	**70**	**232**	**24**	**232**	**24**	0	**194**	**24**	**0.84**
	Total	**1668**	–	**271**	**897**		**897**			**891**	**24**	**0.99**

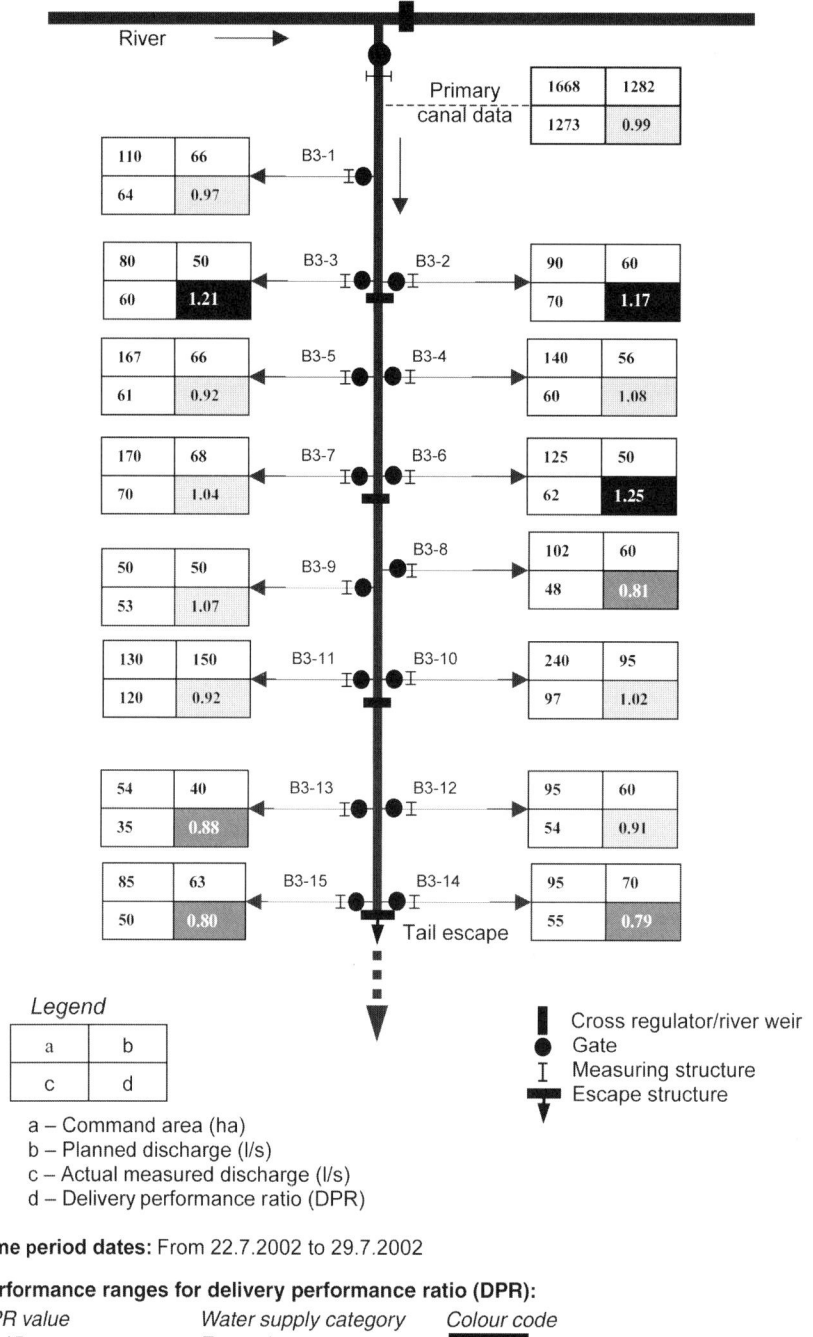

Fig. 4.6. Example of schematic diagram for operational monitoring of delivery performance ratio (DPR) each time period.

5. Reporting results

Reporting results at regular intervals to stakeholders helps to ensure accountability and high levels of performance. This is the point where indicators are useful in providing a communication pathway between service providers, users and other interested parties. The indicators chosen reflect what is relevant for that particular context. The indicators presented in Chapter 3 form a useful base, but managers may also find other types of indicators useful. For example, the number of complaints received may be an indicator of the quality of service. Another example would be the percentage of events when the specifications were not met, or the percentage of time a canal is operational. Financial data and indicators, such as the fee collection ratio, are some of the most important and interesting pieces of information for stakeholders.

Many successful irrigation service providers prepare an annual report where these indicators are presented. These are widely disseminated to users and other stakeholders. This process of reporting also helps to establish accountability, as it provides a check as to whether or not providers are doing their job. An example of a limited set of indicators used annually for monitoring and reporting on the performance of water users' associations is presented in Table 4.6.

Table 4.6. Example of annual performance assessment of WUAs and federations.[a] ISF, irrigation service fee.

No.	Indicator	Definition	Scoring	Score
1.	WUA membership ratio	Total number of WUA members / Total number of irrigators in service area	2 = >50% 1 = 25–50% 0 = <25%	
2.	Annual general meetings	Annual general meeting held	2 = Yes 0 = No	
3.	Annual general meeting attendance	Number of WUA members attending AGM / Total number of WUA members	2 = >50% 1 = 30–50% 0 = <30%	
4.	Administrative council meetings held	Number of meetings held during the year (January–December)	2 = >5 1 = 1–5 0 = 0	
5.	Administrative council elections	Number of elections for members of administrative council held in last 2 years	2 = Yes 0 = No	
6.	Women members of administrative council	Number of women members of administrative council	2 = 1 or more 0 = None	
7.	Employment of accountant	Accountant employed and duration of employment	2 = Yes, >4 months 1 = Yes, <4 months 0 = None	
8.	Area managed by Water Masters	Total gross area serviced by the system / Number of Water Masters employed by WUA	2 = <250 ha 1 = >250 ha 0 = No Water Masters	

Table 4.6. *Continued.*

No.	Indicator	Definition	Scoring	Score
9.	ISF collection per hectare of gross service area (GSA)	$\dfrac{\text{Total ISF collected}}{\text{Total gross area serviced by the system}}$	2 = >25 $/ha 1 = 15–25 $/ha 0 = <15 $/ha	
10.	ISF collection as percent of target	$\dfrac{\text{Total ISF collected}}{\text{Target total annual ISF}}$	2 = >90% 1 = 60–90% 0 = <60%	
11.	ISF collection per hectare irrigated	$\dfrac{\text{Total ISF collected}}{\text{Total annual irrigated crop area}}$	2 = >20 $/ha 1 = 15–20 $/ha 0 = <15 $/ha	
12.	Financial audit of WUA	Level of approval of WUA financial affairs by independent auditors	2 = Accounts approved 1 = No audit undertaken 0 = Accounts qualified/rejected	
13.	Area transferred to WUA	$\dfrac{\text{Area transferred to WUA}}{\text{Total gross area serviced by the system}}$	2 = 100% 1 = 50–99% 0 = <50%	
14.	Annual maintenance planning	Extent of annual maintenance planning, costing and implementation Note: The inspection plan must be reviewed and scored by the monitoring personnel	2 = Inspection undertaken and detailed plan produced 1 = Maintenance plan produced, no proper inspection 0 = No plan produced	
15.	Degree of flow measurement	Level of flow measurement at the head of the system (either primary canal or secondary canals)	2 = Full measurement record 1 = Some water measurement 0 = No measurement	
16.	Maintenance expenditure per unit GSA	$\dfrac{\text{Maintenance cost}}{\text{Total gross area serviced by the system}}$	2 = >15 $/ha 1 = 6–15 $/ha 0 = <6 $/ha	
17.	Maintenance expenditure to revenue ratio	$\dfrac{\text{Maintenance expenditure}}{\text{Gross revenue collected}}$	2 = >70% 1 = 40–70% 0 = <40%	
18.	First irrigation crop area ratio (of GSA)	$\dfrac{\text{Total annual recorded (first) irrigation crop area}}{\text{Total gross area serviced by the system}}$	2 = >50% 1 = 30–50% 0 = <30%	
19.	Crop audit correction factor	$\dfrac{\text{Reported area of first irrigation}}{\text{Crop area measured from crop area audit survey}}$	2 = >90% 1 = 75–90% 0 = <75%	
	WUA total score	Sum of scores for performance indicators	2 = >32 1 = 20–32 0 = <20	

[a] Assessment of the federation is made through analysis of the performance of the individual WUAs making up the federation.

6. Taking action

The most important reason to do the assessment is to take action when needed. When flows are not being delivered according to target, some action is necessary. This may be a simple adjustment, or it may be more complicated, requiring diagnostic analysis (discussed in Chapter 5).

The potential action that can be taken is shown in Fig. 4.7. If operational targets are not met, diagnostic analysis is used to identify the causes and action taken, where feasible, to address these causes. If identified causes for not attaining the operational targets cannot be removed, it may be necessary to alter the target levels in the service agreement. Even if operational targets are met, it is advisable to question whether they require review. An example would be where operational targets are not being met due to low levels of motivation by field staff. The solution might be increased salaries and/or performance-related pay, but because they are in government service salaries are strictly graded, and performance-related pay not acceptable. In such circumstances it may be necessary to downgrade the expectations in the service agreement. A possible feasible solution identified during the diagnostic analysis might be to hand over the system to water users.

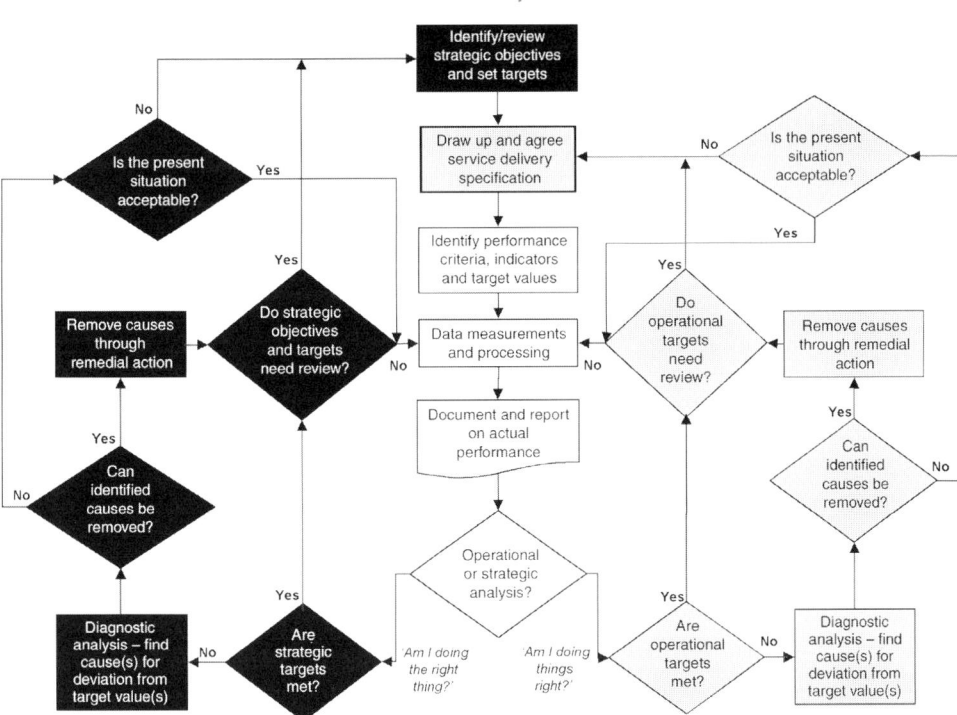

Fig. 4.7. Procedure for taking action following strategic and operational performance assessment.

A similar process can be followed if strategic targets are not met, using diagnostic analysis to understand why performance targets are not being met. If identified causes of low performance cannot be resolved, then the strategic objectives may need to be reviewed. An example would be where the groundwater level is rising to within the root zone of the crop. The diagnostic analysis might identify a number of potential solutions, some of them feasible, some of them not. Buried pipes could be a solution, but might be too expensive, whereas better water management practices might provide a cost-effective and feasible solution.

References

Huppert, W. and Urban, K. (1998) *Analysing Service Provision: Instruments for Development Cooperation Illustrated by Examples from Irrigation.* GTZ publication no. 263. Deutsche Gesellschaft fur Technische Zusammenarbeit (GTZ), GmbH, Eschborn, Germany.

Malano, H.M. and van Hofwegen, P.J.M. (1999) *Management of Irrigation and Drainage Systems – a Service Approach.* IHE monograph no. 3. A.A. Balkema, Rotterdam, The Netherlands.

Replogle, J.A. and Merriam, J.L. (1980) Scheduling and management of irrigation water delivery systems. In: *Irrigation – Challenges of the 80s.* American Society of Agricultural Engineers, Second National Irrigation Symposium, Nebraska, 20–23 October, pp. 112–126.

5 Diagnosing Irrigation Performance

Introduction

Diagnostic assessments are made to gain an understanding of how irrigation functions, to diagnose causes of problems and to identify opportunities for performance gains in order that action can be taken to improve irrigation water management. Diagnostic assessments are carried out when difficult problems are identified through routine monitoring, or when stakeholders are not satisfied with the existing levels of performance achieved and desire a change. Diagnostic assessment supports both operational performance monitoring and strategic planning. Box 5.1 illustrates various purposes and players that may be involved in diagnostic assessments.

Figure 5.1 illustrates the relationship between routine operational irrigation performance monitoring[1] and diagnostic assessment. The left loop represents a normal situation of irrigation operations where plans and procedures are carried out and monitored in periodic day-to-day or year-to-year cycles. Normally when targets are not met, simple adjustments can be made to procedures to bring back the system to normal operations. In some cases, there may be a problem, or a felt need to improve the situation that requires more investigation, and a diagnostic assessment is called for to identify the problem and its causes, and to recommend ways that improvements can be made. Diagnostic assessment differs from routine operational monitoring, in that it is typically an episodic event that can be repeated as and when necessary.

Operations are carried out and monitored with the aid of performance indicators. If there is a problem, if targets are not met, or there is a desire for change, there is a need to jump out of the loop into diagnostics. Within diagnosis, causes of problems and solutions are identified and reported back to management, who then update either strategic or operational plans, and carry on with the normal management cycle.

© M.G. Bos, M.A. Burton and D.J. Molden 2005. *Irrigation and Drainage Performance Assessment: Practical Guidelines* (M.G. Bos et al.)

> **Box 5.1.** Characteristics of diagnostic assessments.
>
> Purposes of diagnostic analysis
> - To diagnose operational problems and suggest solutions.
> - To identify potential changes in support of strategic planning.
>
> For whom?
> - Irrigation and drainage service providers and water managers.
> - Government, lending and funding, and research agencies.
> - Farmers and other service recipients.
>
> From whose viewpoint is diagnostic analysis done?
> - Service providers.
> - Service users.
> - Outside agency.
>
> Who does diagnostic analysis?
> - Irrigation service provider or management agency.
> - Outside team of specialists.

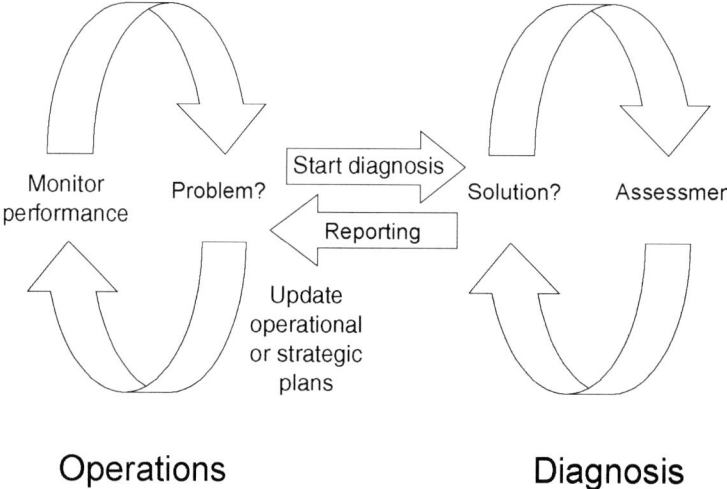

Fig. 5.1. Relationship between the normal operational and strategic management cycles in irrigation.

During operations, operators, managers and users routinely collect and scrutinize information. Performance indicators such as those in Chapter 3 are ideal for routine operational monitoring. One question that constantly arises is whether targets are being met. When targets are not met, routine adjustments are attempted. Two general problems may occur:

1. Operational targets are not met even though every effort is made to do so.

2. Expected results are not achieved in spite of operational targets being met.

The first case relates to operational planning and procedures. For example, suppose high groundwater tables become prevalent in a certain area of the irrigated area. Operational and management staff would try to make operational adjustments to solve the problem. If, however, the problem is not easily solved, a more detailed diagnostic assessment is called for to find the causes of the problem, to report the problem plus possible solutions so that adjustments can be made.

The second case relates to strategic planning. When operational targets are met and desired outcomes are not achieved, more fundamental strategic changes are required. Suppose that target values of operational indicators, including delivery performance ratio, are routinely met, but farmers feel constrained in their crop choice because of the timing and amount of deliveries. Farmers in this case wish more income by switching to higher valued crops. The reasons for not achieving desired outputs could be a change within the system (lower target flow rates), a change external to the system (excessive recharge from a newly built canal), or a change in desires of the irrigation community (growing of higher valued crops). In all of these cases, a more detailed diagnostic assessment of the system is helpful and often essential to understand where opportunities lie to address new requirements.

Ideally, diagnostic assessments should be initiated through routine performance procedures by the irrigation management agency. A major theme of these guidelines is in fact to promote more performance orientation from irrigation managers, and most of the discussion is focused on this. There may be other reasons that trigger diagnostic assessment. A planning or funding agency may want to know more about why a certain level of performance has not been achieved or what investments or changes can be made to improve performance. A team of outside specialists in cooperation with the water service agency then studies the irrigation system and makes recommendations for improvements. Table 5.1 gives possible reasons and examples of diagnostic assessments.

Who does diagnostic analysis? For operational or strategic management of irrigation, irrigation agency staff members should normally carry out performance assessment, including diagnostic assessment. If the problems faced require specialist expertise, or objective assessment, an outside consultant may be called in to perform the analysis. Often a third party is asked to perform an assessment commissioned by a planning or funding agency to understand why investments in irrigation are not paying off, and how they could be improved. An independent research agency or a university may perform a diagnostic study to draw general conclusions about irrigation performance.

Table 5.1. Examples of situations that warrant diagnostic assessment.

Reasons for diagnostic assessment	For whom?	Whose viewpoint?	By whom?
Solve operational problems	Service providers and users	Service providers and users	Irrigation management staff
Identify strategic directions	Service providers and users	Service providers and users	Outside consultant, irrigation management staff
Solve external problem such as pollution caused by irrigation	Outside agency (such as environmental regulating agency), service providers	Outside agency, service providers and users	Outside consultants and irrigation management staff
Implement change at dysfunctional irrigation system	Government or financing agency	Water users, government	Outside agency or consultant

In summary:

- Diagnostic assessments are carried out to find out about problems, to understand constraints; to find opportunities for improvement; or to learn more about successes and failures of irrigation design and management. They will be needed (and repeated) when targets are not met and the reason for not meeting the targets needs to be determined.
- Diagnostic analysis complements routine performance monitoring and evaluation of irrigation systems. For performance-oriented management, it is instigated based on information from routine monitoring when there are deviations from target performance levels, when other problems arise or when there is a need to change performance levels.
- Diagnostics should be performed by system management except in special situations. If specialist know-how is needed for the diagnosis, a specialist can be engaged. If instigated by an outside party, consent of the management agency is required.

Basic Concepts and Principles of Diagnostic Assessment

Here basic concepts are presented for diagnostic assessments. When performing diagnostic assessments, assessors should take a systems approach to understand the problem at hand, the assessment team should be composed of the right mix of disciplines, and assessors should involve stakeholders and utilize indicators in their analysis.

Systems approach

Performance of irrigated agriculture is influenced by the interaction of complex human, physical and socio-economic systems (see also Fig. 2.2). For example, farmer adoption of more precise irrigation application methods may be more a function of the economics rather than of lack of knowledge or inappropriate canal deliveries. A system transforms inputs into outputs through various processes. For example, water is an input to a conveyance system, and delivery of water to various farms is an output. There are several processes involved in water delivery, such as planning and allocating water, maintaining infrastructure, conveying the water through a network of channels or pipes, and regulating and measuring flow. Many of these, such as water allocation, are outputs of other systems, in this case a social organizational system. Understanding system process, key linkages between systems and the environment in which systems work is essential to appreciate cause and effect relationships and thus diagnosis.

Need for a variety of perspectives

Irrigation performance is a function of many technical, physical, social and economic processes. It is a function of the design of the system, how the system is operated, who operates the system, incentives for farmers, government policies and institutions for managing water. Explanations drawn solely from any single discipline do not suffice. Understanding performance clearly requires the need for understanding from a variety of perspectives. For solving complex problems, a team with different backgrounds is required. This will be reflected in the performance indicators chosen as discussed in Chapter 3.

The need for user involvement

Diagnostic process should include the stakeholders who will ultimately be influenced by the results of the assessment. If, for example, an irrigation management agency is assessing means to improve delivery service, they need to include the irrigation users within the assessment. Typical stakeholder groups include farmers, irrigation agency, planners and policy makers, and water users not in the irrigation system but immediately affected by irrigation. If outsiders do the assessment, they need to closely involve interested parties, including farmers, irrigation agency and related support agencies. User involvement helps everyone understand the functioning of the system, allows for a variety of ideas and enhances ownership in results and recommendations.

Use and interpretation of indicators in diagnostic assessment

Performance indicators are vital to diagnostic assessments to understand relationships and to develop performance statements about irrigation. The use and interpretation vary between an operational assessment and a diagnostic assessment. For operations, management-set targets are a reference point and performance assessed against these targets. In the diagnostic process, understanding of key processes is crucial and indicators can help in the understanding and relationship between key processes. The indicators presented in Chapter 3 are useful in diagnosis as well as routine operations, but other indicators can be used to complement these. Figure 5.2 illustrates, for example, the interaction between the depleted fraction and the evaporative fraction for the Fayoum depression, Egypt (Bos and Bastiaanssen, 2004). In this context it should be noted that these two fractions also influence the availability of water for leaching and crop yield.

One advantage in using the standard indicators of Chapter 3 is that it allows for comparison across time and other irrigated areas. It is often quite useful to see the value of an indicator relative to another location.

Steps for Diagnostic Analysis

There are two common approaches to understand system performance and diagnose problems. The first approach is to collect as much informa-

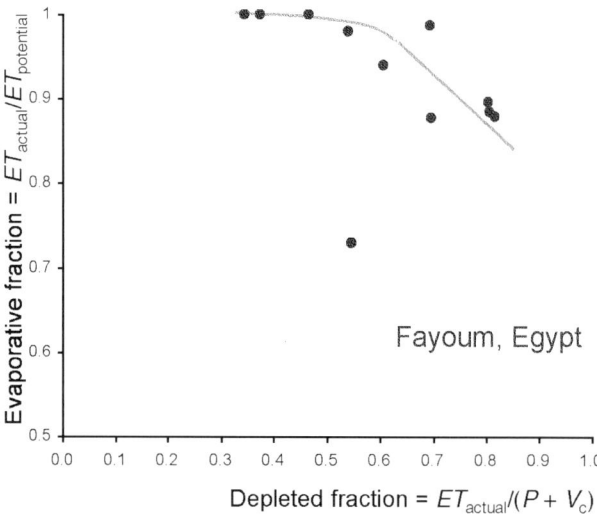

Fig. 5.2. This illustrates how indicators can be used to help enhance understanding of key relationships with diagnostics. In this case, the evaporative fraction begins to decrease at values of depleted fraction above 0.6. Yield levels are likely to decrease at higher depleted fractions, because of the reduction in the ratio of actual to potential evapotranspiration.

tion as possible about the system and explain the functioning of the system through analysis. The second approach is to focus on and trace key cause–effect relationships. While the first approach can yield a broad understanding of irrigated agriculture, it is often expensive to collect, measure and handle data on performance, and is one reason why irrigation managers do not routinely do performance assessment.

An effective means of diagnosing performance adopting the second approach is to take a broad view of the system to identify possible explanations for the problem, then, through successive approximation, offer explanations of reasons for high or low performance. The general approach is to state the problem that is being addressed, make an initial explanation or hypothesis of cause–effect relationships, collect and analyse diagnostic information, then reformulate and re-test the hypotheses until satisfactory answers are found. When they are found, recommendations are formulated and reported to concerned parties. All these steps are done in light of the key principles presented earlier. A six-step approach is framed in Fig. 5.3.

1. Identify initial diagnostic problems or questions

This identification comes directly from the reason that a diagnostic assessment was triggered. Typical types of questions could be: What are

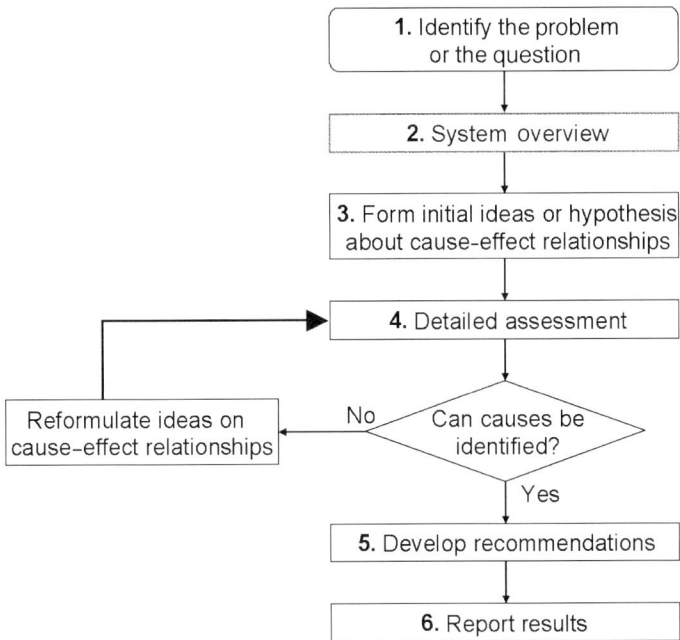

Fig. 5.3. Steps to diagnostic assessments.

the water-related constraints to on-farm productivity? Why is there a head–tail-end problem? Why do tail-enders not pay fees? How can overall productivity be improved? How can problems of water-logging and salinity be reduced? Involve key stakeholders in forming the right set of questions. The question then guides the type of analysis to be done.

2. System overview

To develop an initial hypothesis, take a macro point of view to form a profile of the system of interest, collecting as much related secondary information as possible. When the irrigation management agency is doing the diagnostics, this step may be quickly done. Outsiders especially should first consult local actors knowledgeable about the system. Describe the environment of the system: the climate, soils, water supply and institutional framework using the system characteristics presented in Chapter 2, Table 2.3. The indicators presented in Chapter 3 ensure that information across disciplines is collected. The purpose of the overview is to establish a systems perspective and key relationships that are important for developing an initial hypothesis.

3. Develop working hypotheses

Based on the overview, develop an initial hypothesis about the cause–effect relationships that occur within the system. Use rapid assessment or participatory appraisal techniques to gather enough information to develop initial hypotheses on the behaviour of the system. At this point, take into consideration a range of information from various disciplines and from interaction with stakeholders. It is important to understand the inter-relationships between infrastructure, operations, organizations, economic incentives and policies. Develop working hypotheses, and also a programme to test them. A diagnostic tree presented later in the chapter is useful to understand cause–effect relationships (Fig. 5.7).

4. Detailed assessment

Develop a data collection programme to test the hypothesis, working closely with concerned stakeholders. By first framing the questions and hypotheses, assessors can 'zoom in' on the problem. Certain parts of the system can be studied in more detail, while less information is required from other parts of the system. This minimizes data requirement and streamlines time and effort. Chapter 6 gives more detailed information about data management. Figure 5.4 provides an example of data collected to understand the relationship between groundwater levels and depleted fraction. Figure 5.5 provides an example of operation and maintenance (O&M) budget and fee collection.

During the data collection programme, analyse and interpret results. Does the evidence strongly support the hypotheses? Do stakeholders concur with the conclusions? If yes, the causes of the problems are likely to have been identified. If the hypotheses are not supported, then revisions are required. In this case, reformulate the hypotheses, and develop further testing.

5. Develop recommendations

After finding the cause of the problem, irrigation managers can take action using their regular procedures. Where the clients are the water management agency and water users, it is very important to obtain early feedback. They can be insightful and essential in forming hypotheses, interpreting results and setting up and monitoring a data collection programme. In addition, their early participation allows results to be easily understood, interpreted and used for action.

Recommended actions may include adjustment in operations, or could be concerned with strategic decisions that may be related to routine operations, or may be much more strategic in nature where target

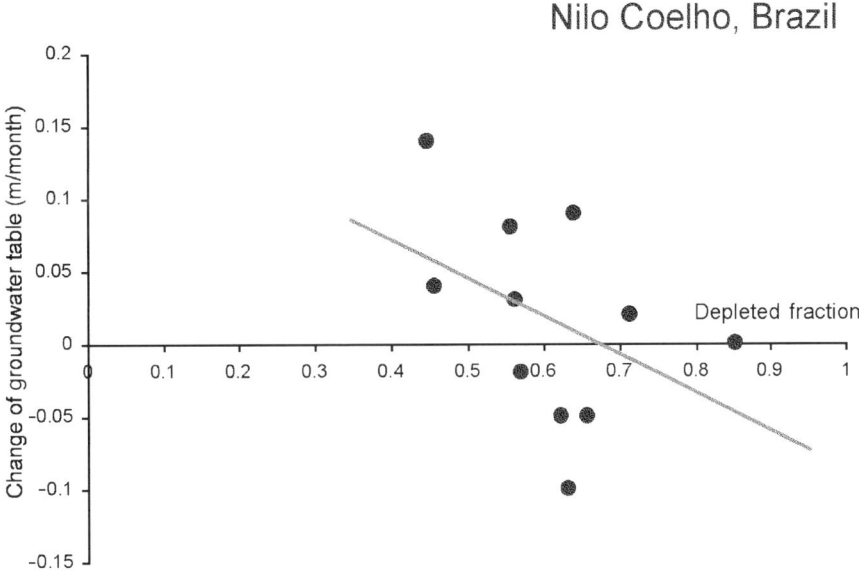

Fig. 5.4. In this case it was hypothesized that decline or accretion of groundwater is related to the depleted fraction. Data were collected from the field and through remote sensing to quantify groundwater levels and depleted fraction. In this case it was shown that at higher levels of depleted fraction, the groundwater levels decline. This information can be used to make plans to stabilize groundwater levels (Bos and Bastiaanssen, 2004).

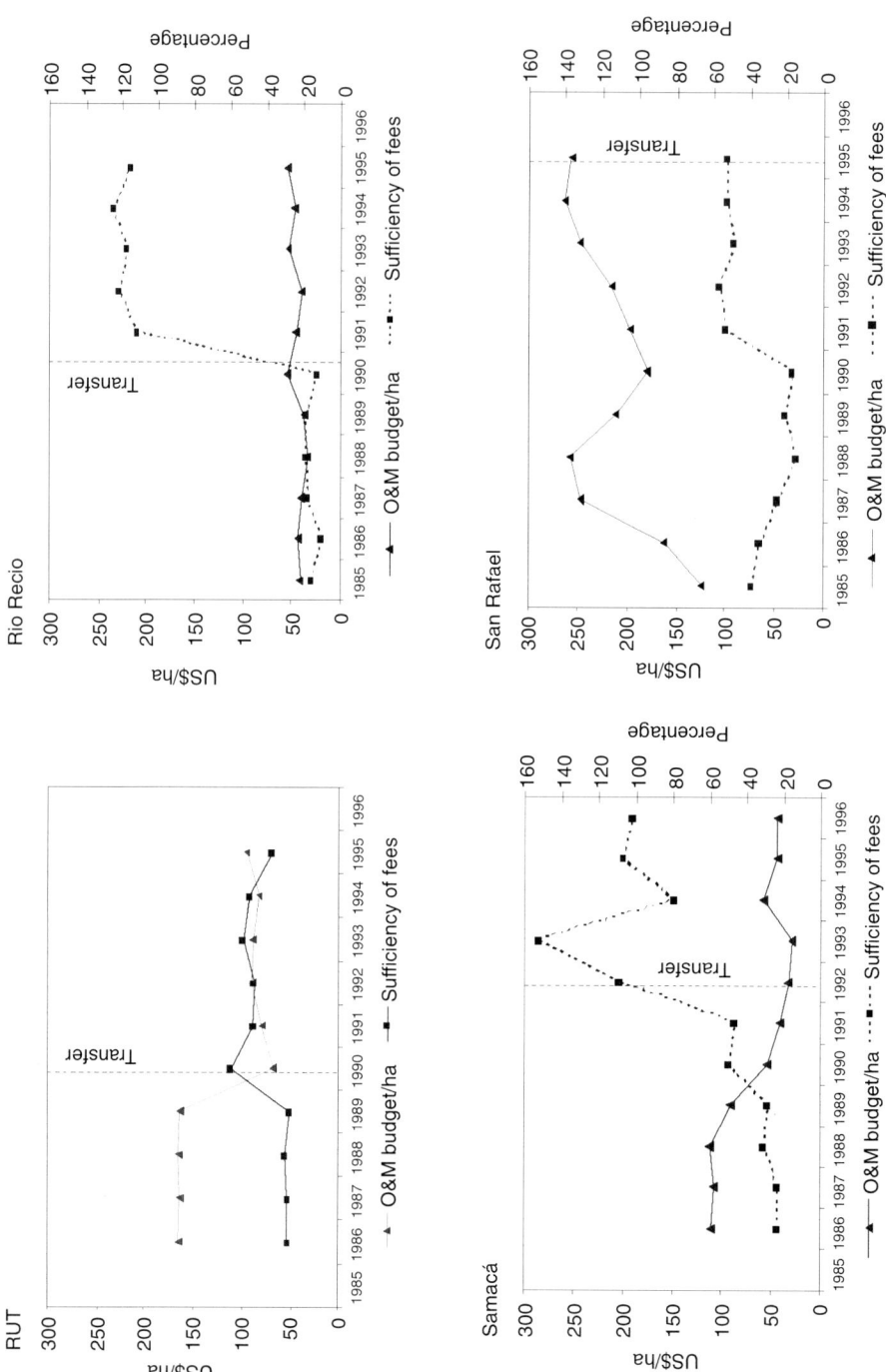

Fig. 5.5. O&M budgets and financial self-sufficiency from four different systems in Colombia. In three out of four systems, fees are sufficient to cover projected budgets (Vermillion and Garcés-Restrepo, 1998).

values of indicators need to be reformulated or new indicators designed. Recommendations should take into consideration what is socially, technically, economically and politically feasible by concerned stakeholders based on information derived from steps 1–4. A performance improvement capacity audit developed by Merrey *et al.* (1995) can be useful to determine whether recommendations are likely to be implemented.

6. Report results

The lead agency doing the assessment should clearly report results to clients. For example, if the management agency does the assessment, a special report should be made for a board of directors and farmers. Reports should clearly state the programme of work done, explanations of key cause and effect relations, the explanation of any problem, supporting data and information supported by tables and graphics, and recommended actions. Reports should be written bearing in mind the key audience for the report.

Methodologies

There are many existing methodologies and analytic techniques applicable to diagnostic assessment. A good assessor will choose the appropriate methodology for the task at hand. The purpose of this section is to give an overview of methodologies that are useful in diagnostic assessment. Several methodologies are described with guidance on when they may be useful, and references are included in the Further Reading section for additional information. The first is on diagnostic methodologies relating to broader approaches or processes applicable to diagnostic assessment, the second part relating to specific techniques useful in diagnosing a particular type or range of problems.

Diagnostic methodologies

Diagnostic analysis methodology

A specific methodology for diagnostic analysis (DA) for assessing and understanding the performance of an irrigated agricultural system has evolved since the 1980s and has been applied to many irrigated areas (Lowdermilk *et al.*, 1983; Clyma and Lowdermilk, 1988; Dedrick *et al.*, 2000). The methodology entails a five-step process, including:

1. Development of an overall plan.
2. Entry – information input.

3. Rapid diagnosis.
4. Detailed diagnosis.
5. Exit – reporting and debriefing.

Key concepts include use of systems analysis to understand complex physical, biological and human systems; interdisciplinary teams to carry out the assessment; action research aimed at implementing organizational change; and user involvement throughout the process. Many of these DA concepts have been important in developing this chapter. DA is conceived in an overall management improvement process aimed at management planning and improving irrigation performance. Typically, DA is aimed at an overall understanding of the system rather than solving a single problem, or understanding performance of a single aspect of a system such as water delivery or economics.

The DA process has been carried out at numerous irrigation systems (as examples see: Alwis et al., 1983; Laitos et al., 1985; Dedrick et al., 2000). Typically, DA studies were initiated by agencies other than the irrigation management agency, such as other government agencies and donor agencies, but the involvement and agreement of the management agency is essential. The DA is taken from a variety of viewpoints, including the farmer's, the irrigation manager's and society's. The DA methodology has always been carried out by an outside team of interdisciplinary researchers as opposed to irrigation agency staff (although they are involved). The DA methodology seems well suited to support strategic planning, but not for solving specific problems. This methodology requires intensive use of human and financial resources, but the payback in terms of improved irrigation performance is expected to be high. The experience and examples of DA have yielded a variety of specific methodologies crossing disciplines that are quite useful within and outside the context of DA (Podmore and Eynon, 1984; Oad and McCornick, 1989).

Rapid appraisal

Rapid appraisal, as the name suggests, is used to give a quick overview of system performance. This is typically used in the initial steps of performing diagnostic analysis. As a result of a rapid appraisal, an initial hypothesis can be developed. At times, an overview based on a rapid appraisal can shed sufficient light on an irrigated area for decisions to be made.

Rapid appraisal techniques rely on field observations plus the collection and review of available data and information. The following sources of information are useful: review of secondary data, interviews with individuals and groups, and observations of various parts of the system. A checklist of information is often helpful to provide an initial guide for the rapid appraisal (see Box 5.2 for an example checklist). Surprising pieces of information may appear and should not be discarded. A rapid appraisal should provide key information to form a profile of the system, information on a few key indicators and other explanatory information to form the basis for key hypotheses. Rapid appraisals can sometimes quickly trace the origin of malfunction, allowing for application of corrective actions and sometimes eliminating the need for a detailed diagnostic analysis.

The advantages of rapid appraisal lie in the ability to quickly form an idea about the system's functioning. Rapid appraisal can point swiftly to the origin of the malfunction, allowing for rapid corrective action, and minimizing the time and effort for detailed diagnostics. The disadvantages are that it relies on the skills of the assessor, many of the results are subjective and potentially misleading, and it is difficult to yield conclusive evidence through this technique.

Participatory rural appraisal

Locally, irrigation communities possess tremendous knowledge about the operation and performance of irrigation. This is an extremely valuable source of information, even for irrigation management agencies, in assessing irrigation performance. Participatory rural appraisal relies on local knowledge to identify problems and develop interventions.

Participatory rural appraisal (PRA) is a family of approaches and methods to enable local people to share, enhance and analyse their knowledge of life and conditions, and to plan and act (Chambers, 1994). PRA is related to and evolved from the rapid rural appraisal techniques (Chambers and Carruthers, 1985; Yoder and Martin, 1985; Pradhan *et al.*, 1988; Grosselink and Thompson, 1997). The local community participates in the research by developing sketches and maps, transects showing resource use patterns, seasonal calendars, trend analysis and daily activity profiles.

Through PRA, information that would have otherwise gone unnoticed is tapped. By involving stakeholders in research and development, there is more likelihood of better acceptance of interventions. A disadvantage is that the quantitative base of information may be weak. For example, this would not be used to generate data on water resources, although it could be helpful in developing a feel for the magnitude of flows when data are missing. While it is an excellent tool for deriving local knowledge, placing this knowledge in the context of broader issues such as basin-wide water use may be missing. Similar to the rapid appraisal techniques, this technique also relies heavily on the skills of the assessor. PRA can be an excellent complement to other tools when assessing performance. PRA techniques are ideally suited for developing and improving service arrangements between the providers and users. For diagnosis, PRA can be used both in initial screening and for a more detailed data collection.

Diagnostic tree

To find solutions to the observed field problems, causes of the problems need to be identified and addressed. Typical irrigation problems have several possible causes (Fig. 5.6), so the task of diagnosis is to sort out

Box 5.2 Rapid appraisal checklist for assessing the performance of a water users' association in an irrigation management transfer setting (extracted from Vermillion et al., 1996).

1. WUA Assessment

1.1 When was the organization established and by what authority?

Month/year: _____

Authority: _____

Month/year of actual transfer of operations: _____/_____

Month/year of actual transfer of maintenance: _____/_____

Month/year of actual transfer of financing: _____/_____

Month/year of official transfer: _____/_____

1.2 In which of the following functions is the WUA involved and for which does it have full authority for policy, planning and implementation? (check all that apply)

Functions	WUA involved?	Full authority?
Irrigation operations within WUA area		
Irrigation operations on main system		
Irrigation maintenance within WUA area		
Irrigation maintenance of main system		
Financing irrigation costs		
Selection of WUA leadership		
Settling disputes between farmers		
Obtaining/arranging agricultural credit		
Obtaining/disbursing agricultural inputs		
Crop processing and marketing		
Sideline businesses		
Other (specify)		

1.3 Legal and political powers of WUA

Function	Formal legal authority?*	Recognized by government authority?*
Can negotiate a water right at level of the WUA (in volume, duration of flow, or share)		
Can define water rights or allocation rules for individual farmers (in volume, duration of flow or share)		
Can enforce fee collection from farmers		
Can fine farmers in cash or in kind for rule violation		
Can withhold irrigation water from farmers for rule violation		
Can seize land ownership from farmers for rule violation		
Can take farmers to court of law for rule violation		
Has rights-of-way for irrigation structures		
Owns irrigation infrastructure		
Can own property		
Can have bank account		
Can obtain and extend credit		
Can make business and employment contracts		
Other (specify)		
Other (specify)		

*In columns for legal authority and recognition by local government, enter 1 = full, 2 = partial, 3 = no, 4 = unclear.

Diagnosing Irrigation Performance

1.4 Which of the following are required in order to be a member of the WUA? (check all that apply)
    1. Must own land in the ICA of WUA ❏
    2. Must be a primary cultivator in WUA irrigated area ❏
    3. Must own water right or share in WUA irrigated area ❏
    4. Must register and pay membership fee ❏
    5. Must be only household member representing the farm ❏
    6. Other (specify) ❏

1.5 Which of the following are grounds for losing membership in the WUA? (check all that apply)
    1. Failure to pay annual fees ❏ 2. Failure to support WUA maintenance activities ❏
    3. Violation of rules ❏ 4 Farm owner disposes of farm in WUA area ❏
    5. Farm owner ceases to cultivate farm in WUA area ❏ 6. Non-owner cultivator ceases to cultivate in WUA area ❏
    7. Member alienates own water right ❏ 8. Other (explain) ❏

1.6 WUA leadership and staff positions

Leadership position	Function	No. of persons M F	How selected*	Full/ part-time	How paid**

Staff position	Function	No. of persons M F	How selected*	Full/ part-time	How paid**

*Codes for "How selected": By govt. official = 1, by farmers = 2, by both govt. officials and farmers = 3, by volunteering = 4, hired by WUA leaders = 5, Other = 6.

**Codes for "How paid" (select all that apply): Salary/honorarium = 1, unofficial compensation = 2, share payments from farmers = 3, income from WUA business activities = 4, not paid = 5, other = 6.

1.7 In which of the following have women ever been involved in the WUA? (check all that apply)

    Held WUA leadership positions ❏   Held WUA staff positions ❏
    Attended WUA meetings ❏   Spoken at WUA meetings ❏
    Personal lobbying of WUA ❏   Joined WUA maintenance activities ❏
    Paid irrigation fees ❏   Assisted with bookkeeping ❏
    Assisted with operations ❏   Assisted in settling disputes ❏
    Other (specify) ❏

1.8 What is the most severe punishment, fine or penalty which has been applied against a farmer(s) by the WUA since transfer? What was the infraction by the farmer(s)?
Punishment: _____

Infraction: _____

which cause is leading to the undesirable effect. Tracing cause and effect relationships gives rise to a diagnostic tree, a powerful tool for identification of the root cause of a given problem (Kivumbi, 1999; Kivumbi et al., 1999).

The problem and identified possible causes shown in Fig. 5.6 can be structured into a hierarchical diagnostic tree, as shown in Fig. 5.7. Each of the possible causes can be investigated at the same time, or a scoping study or rapid assessment carried out and the most likely cause investigated first, followed by the next most likely if the first is not successful in identifying the root cause. It is important to note that a cause can also be a problem, as shown in Fig. 5.7.

It is also important to note that whilst it is important to trace the root cause, there may be intermediate solutions along the way. For example, the sediment in the canal could be removed. This solution would help in the short term, but might not be cost-effective if the root cause was the damaged control structure at the river intake, allowing sediment to enter the canal during flood periods.

Additionally, whilst problems might be identified, solutions might not be possible. A common problem with management of government agency-run irrigation and drainage systems is the relatively low salaries paid to O&M staff, and the inability to offer, within the government pay structure, performance incentives. Whilst a pay scale that is performance related and motivates O&M staff might be identified as a solution to operational problems, it may not be feasible under government regulations. In some cases the solution that is being taken in this case is to transfer management to water users' associations.

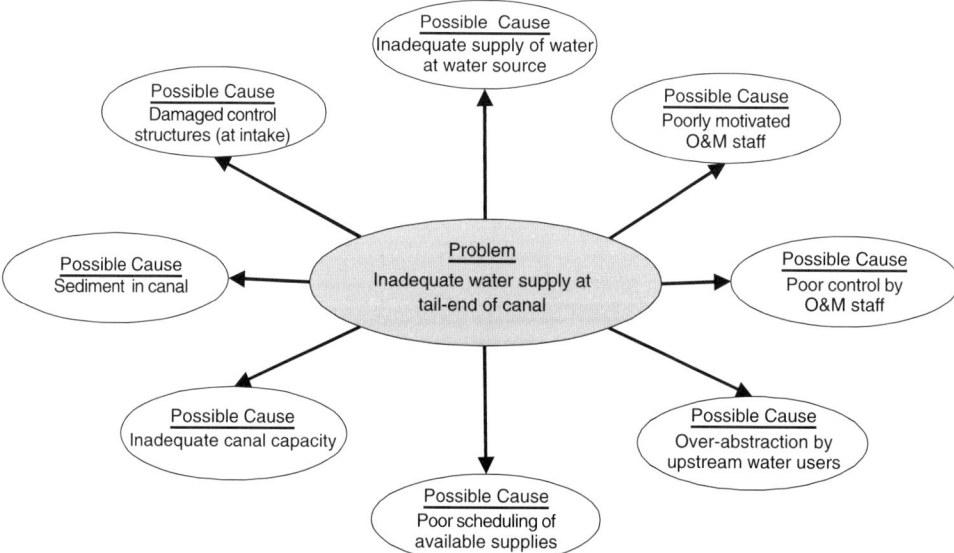

Fig. 5.6. Problem with possible causes.

Diagnosing Irrigation Performance

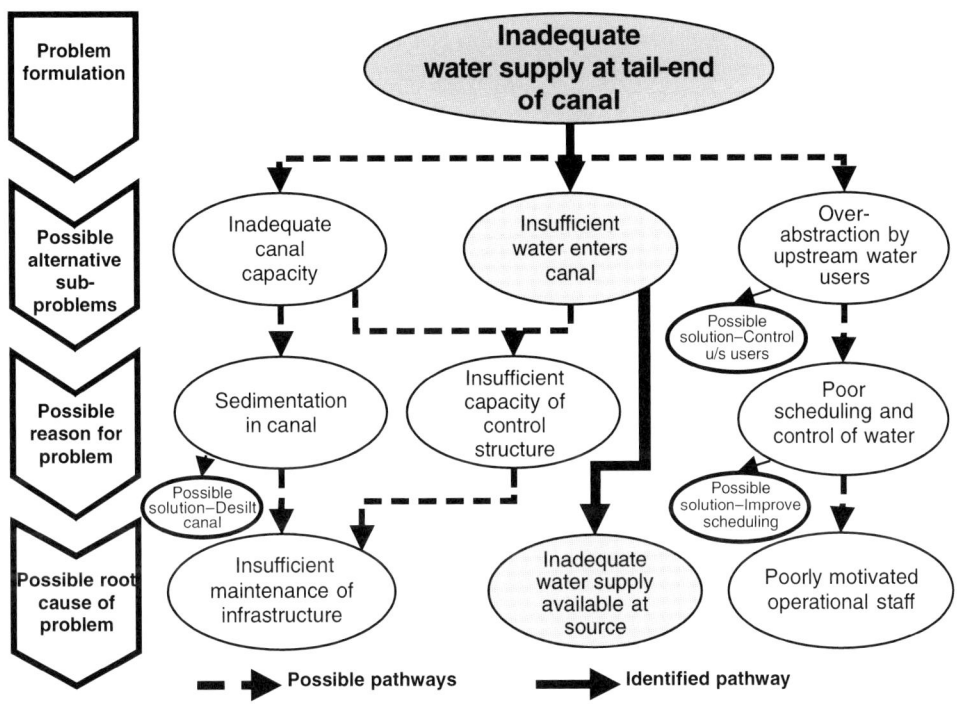

Fig. 5.7. Structuring the problem and likely causes into a diagnostic tree.

Specific Diagnostic Assessment Techniques

Remote sensing

Remote sensing techniques are increasingly being utilized in performance assessment and are in many situations quite useful for diagnostic assessments. The use of remote sensing has several distinct advantages over traditional ground data collection. Remote sensing can be used to gather information over an entire area, while ground data collection relies on sample areas. Data collection by remote sensing does not intrude into the day-to-day life of those in the irrigation community. Often, the presence of observers changes the behaviour of those being observed, so the information collected does not reflect normal operating conditions. Data can be disaggregated to the resolution of the image, or aggregated up to useful units such as various service areas within an irrigation system. Because satellite images have been available since 1982, development trends can be established looking 20 years back.

Remote sensing can yield key information with sufficient accuracy, as shown in Table 5.2. An example of remote sensing results from the Bhakra irrigation system in India is presented in Fig. 5.8.

Table 5.2. Accuracy of remote sensing parameters in drainage estimated by the Ede-Wageningen Expert Consultation at a 95% confidence interval. The range of accuracy indicated by individual scientists is added (Bos et al., 2001).

Thematic parameter	Accuracy (%)	Standard deviation accuracy assessment (%)	Coefficient of variation accuracy assessment (%)
Topographical characteristics	81	16	20
Land use	84	8	10
Land wetness (drought index)	78	8	10
Soil moisture (surface)	70	11	16
Soil moisture (root zone)	64	25	39
Waterlogging	87	7	8
Drainage from area	78	3	4
Salinity occurrence on surface	77	19	25
Soil salinity	63	18	28
Irrigated area	85	8	10
Crop identification	78	12	16
Reference ET	81	12	14
Potential ET (crop coefficient)	79	8	10
Actual ET	83	9	10
Leaf area index, LAI	80	9	11
Biomass growth	79	12	15
Crop yield	72	19	19
Water rights	93	3	3
Soil erosion	68	38	38
Average	78		

The cost of obtaining remotely sensed data is often cited as a constraint to its use. Prices are decreasing rapidly, and the quality and resolution of images are improving. For certain types of data like irrigated area, or land-use cover, the cost of data collection is less than 25% of conventional data collection programmes. Remote sensing cannot substitute for local field-level knowledge and experience, and is applicable to a limited set of problems that may occur.

Gender performance indicators for irrigation

The different roles of women and men in water management, food production, marketing and household consumption influence performance of irrigation. Performance assessment, policies and interventions have been poor in consideration of these differing roles. The Dublin Principles, which emphasize the role of women in water management and the growing and widely recognized trend of 'feminization of agriculture' underscore the importance of gender.

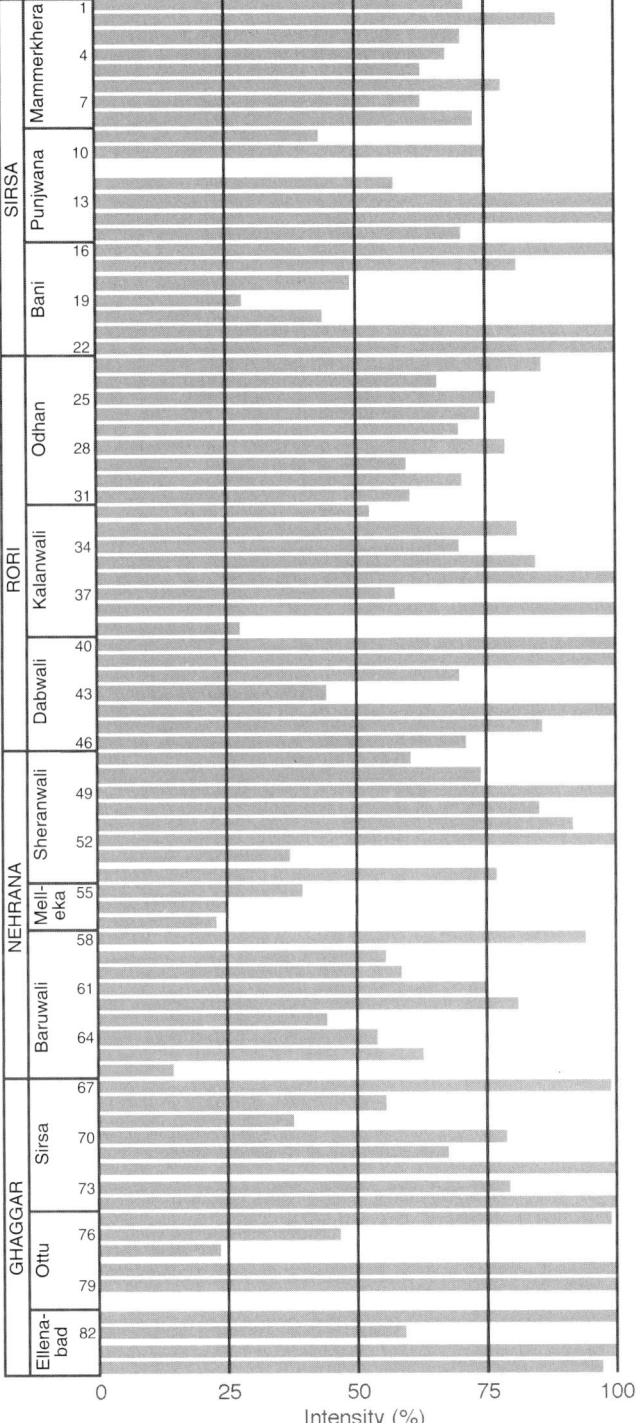

Fig. 5.8. Irrigation intensity in administrative command areas of the Sirsa Irrigation Circle, rabi 1995/96, from Bastiaanssen *et al.* (1999). Remote sensing in this case is useful for obtaining a disaggregated picture of irrigation.

A major reason for this gap is a lack of understanding of gender–water–food-security issues (Merrey and Baviskar, 1998). Part of the difficulty lies in the tremendous global variation in the gendered organization of water management. It is helpful to classify farming systems into female-managed, male-managed and mixed, depending on who is primarily taking decisions in agriculture (Van Koppen, 2002). Female-managed farming systems, for example, are common in much of sub-Saharan Africa. With this initial recognition, much can be done to design better interventions and support systems.

Gender performance is indicated by a qualitative statement on men's and women's inclusion on a range of issues, including land and water rights, membership in various fora and leadership. An example from the rice growing area of Comoe province of Burkina Faso is illustrative. In this area, 80–90% of the farmers are women. Table 5.3 shows the gender performance before and after project intervention. As part of the initial project intervention, land rights were expropriated from women farmers, thus the gender performance diminished. Later projects took into consideration gender differences, recognizing and organizing female and male farm decision makers before project construction, with improved results as shown in Table 5.3. It was reported that maintenance conditions and productivity were better in the projects implemented with better gender performance.

Water balance and accounting

Establishing a water balance is an essential step in the identification of opportunities for water savings or in increasing the productivity of water. This is especially true in environments with high competition for water, and where irrigators are under pressure to save water that can be used for the environment, cities, industries or for more agriculture. The

Table 5.3. The example illustrates the use of the gender performance indicator. Performance is rated by a plus or minus symbol depending on inclusion of women and men under each category. In this case, gender performance declined after the first ill-conceived projects were constructed, but improved after improvements in project design were made.

Actor	Before project Local arrangements	First two trials Agency	After improvement Local, then OK by agency
Land rights	+	–	+
Membership rights	+	–	+
Water rights	+	+	+
Inclusion into fora	+	–	+
Inclusion as leaders	+/–	–	+
Function as leaders	+/–	–	+/–

general procedure is to perform a water balance, then classify various water balance components into various categories to understand in which category opportunities for improvement lie. Water balance studies rely on a mass balance equation for a defined domain bounded in time and space, where inflow into the domain balances with outflow plus changes of storage within the domain. Inflows include rainfall, groundwater and surface water flows into the domain. Outflows are evaporative fluxes across the top layer of the domain plus any lateral surface or subsurface flows. Changes in storage include changes of volumes in the soil profile, the groundwater system and surface reservoirs.

Water balance studies at irrigated areas have proven useful to understand flow paths of water, to quantify efficiency and to better understand beneficial and reasonable use (Bos and Nugteren, 1974; Kijne, 1996; Perry, 1996; Tuong et al., 1996; Burt et al., 1997). Water accounting classifies inflows and outflows into categories of use (Molden, 1997; Molden and Sakthivadivel, 1999). Some uses of water are beneficial, others non-beneficial. Some uses are intended for a certain process such as agriculture (crop ET), urban (drinking and cleaning) or industry. Some uses are unintended (irrigation water used as a drinking water source) or natural (evaporation by trees, percolation). Water accounting allows us to track how much water is diverted to and depleted by each use, and it gives an estimate of the remaining water for further development. It is useful to determine how much water can be saved, reused by other users downstream or reallocated to more productive uses. To illustrate, examples from Chishtian, Pakistan (Molden et al., 2000) and Kirindi Oya, Sri Lanka (Renault et al., 2000) are used.

The Chishtian irrigated area, with a command area of 71,000 ha, is located in Pakistan's Punjab with a landscape heavily dominated by agriculture. Water accounting studies were performed to assess use and productivity of water (Molden et al., 2000). During the 1993/94 agricultural year, 670 million cubic metres (MCM) of water entered the boundaries of the water balance domain[2] from irrigation deliveries, rain and subsurface flows (Fig. 5.9). Total depletion comprised 600 MCM in the form of crop evapotranspiration, and an additional 30 MCM was estimated as evaporation from municipal and industrial use. The ratio of human depletion to inflow (depleted fraction = 630/670 = 0.94) is already quite high. In addition, it was estimated that an additional 100 MCM was evaporated by home gardens, forests and other land surfaces, and that there was a net removal from groundwater of 70 MCM to attain the balance.

From this perspective, farmers are very effective in converting water into crop production. But, groundwater was mined during the year, and in this area very little water was available for environmental purposes such as flushing salts or for ecosystem sustenance. The study indicates that there is little scope for water savings at the present level of crop evapotranspiration, and potentially a situation of non-sustainable use. To verify this, indicators on groundwater from Chapter 3 should be

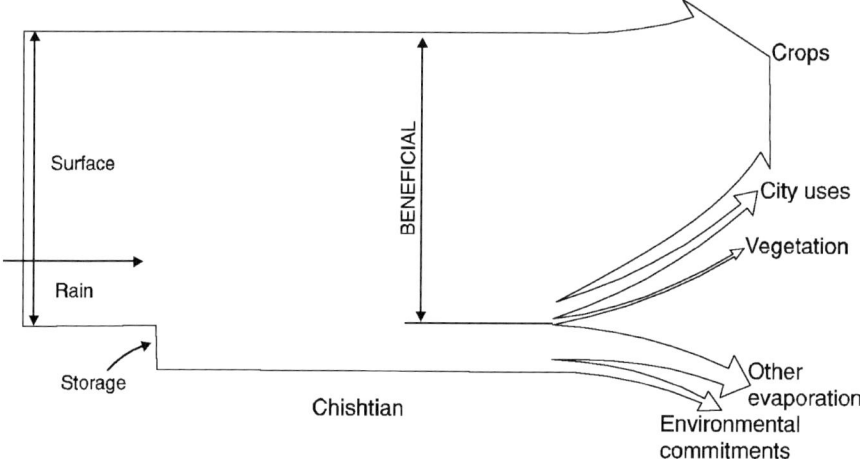

Fig. 5.9. Water accounting diagram from the Chishtian Irrigated Area in Pakistan. The width of the areas are in proportion to the water inflow or depletion. In this case water is removed from groundwater storage to meet crop, city and other evaporative demands, plus environmental commitments. Farmers as a group, because of pumping of return flows to groundwater, are very effective in converting water inflows to beneficial use.

carefully monitored over time. Solutions should be targeted at either reducing evaporation or less beneficial ET, increasing inflows (which is unlikely) and increasing productivity of water consumed by evapotranspiration.

As a second example, the Kirindi Oya system, serving a command area of 8600 ha in southern Sri Lanka, shows another pattern of water use (Table 5.4). In this case (Renault *et al.*, 2000), 475 MCM entered the system as irrigation diversions (245 MCM) and rain (230 MCM). Crop evapotranspiration was estimated at 95 MCM, while surprisingly evapotranspiration from trees and home gardens was 180 MCM. A measured value of 96 MCM was drained by irrigation out of the area, but part of this amount was necessary for downstream lagoons and fisheries. In this case, the process fraction when only considering crops was (95/475 = 0.20), but when considering other clearly beneficial uses such as trees (like coconut and mango trees) the fraction beneficially depleted was (280/475 = 0.50). The case illustrates that there is scope for water savings by reducing drainage outflows. But it also indicates the need to carefully account for environmental needs, and the other beneficial uses before proposing appropriate changes.

Questionnaire surveys

Questionnaire surveys play an important role in a diagnostic assessment of irrigation. From questionnaires, important information on yields, agri-

Table 5.4. Water Accounting Table for Kirindi Oya irrigation system from Renault et al. (2001). Inflow and outflow numbers were obtained from a variety of flow measuring and estimating procedures. Uncommitted outflows in excess of downstream environmental requirements represent scope for additional water savings. Non-process but beneficial depletion was high in comparison to crops and provides additional information on the performance of irrigation, in that when these are considered, beneficial use of water is much greater than if crops are considered alone.

Components	Total	Parts	As % of available water
Gross inflow	475		
Irrigation releases from reservoir (measured)		245	
Precipitation on water balance domain (measured)		230	
Subsurface inflow (estimated)		0	
Storage change	3		
Surface storage (measured)		3	
Subsurface storage (estimated)		0	
Net inflow	478		
Committed outflow (nominal amount, estimated based on environmental needs or downstream rights or requirements)	47		
Available water for use within domain	431		
Depletion			
Process			
Irrigation – Crop ET (measured and estimated)		96	22%
M&I (estimated)		negligible	
Non-process, non-beneficial or low valued			
ET of water bodies and fallow land (measured and estimated)		56	13%
Non-process, beneficial			
Home gardens, forest		184	43%
Outflow (measured)	143		
Uncommitted outflow		96	22%
Committed outflow (to meet needs of downstream wetlands)		47	

cultural input use, social organization and conflicts can be obtained. Questionnaires can be developed by interdisciplinary teams to obtain a range of input. A well-designed survey based on statistical methods can provide a strong quantitative base from which decisions can be made. The literature on theory and application of surveys is quite large, and a short list of references is provided at the end of the chapter.

Figure 5.10 shows the results from a questionnaire aimed at gaining insight into water delivery performance before and after an intervention. These qualitative results can be used to complement more quantitative measures to develop more effective delivery programmes.

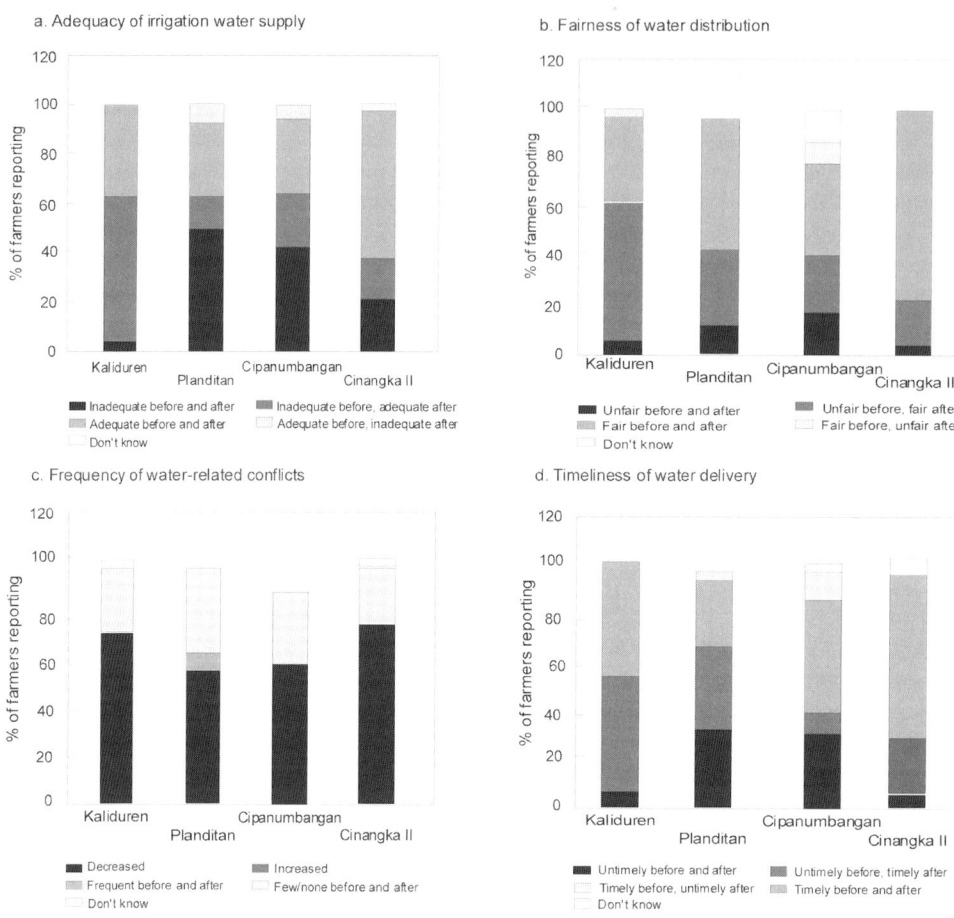

Fig. 5.10. Farmer perceptions about operational performance from four irrigation systems in Indonesia derived from a questionnaire survey (Vermillion *et al.*, 2000).

Notes

[1] Operational performance monitoring relates to day-to-day, season-to-season monitoring and evaluation of performance.

[2] 504 MCM from irrigation diversions, 143 MCM as rain and 73 MCM as net groundwater abstraction. Crop evapotranspiration was 595 MCM, while evaporation from cities was about 50 MCM.

References

Alwis, J., Nelson, L., Gamage, H., Nandasena, R.A., Griffin, R.E., Yoo, K., Ekanayake, A., Haider, M., Wickramasinghe, L., Dunn, L., Bondaranaike, M.A.W., Reddy, J.M. and Laitos, W.R. (1983) *Diagnostic analysis of farm irrigation systems on the H-Area of Kalawewa in the Mahaweli Development*

Project, Sri Lanka. Draft report prepared in cooperation with USAID (contract AID/DSAN-C-0058) x, 172 pp. (WMS report no. 16).

Bastiaanssen, W.G.M., Molden, D.J., Thiruvengadachari, S., Smit, A.A.M.F.R., Mutuwatte, L. and Jayasinghe, G. (1999) *Remote Sensing and Hydrologic Models for Performance Assessment in Sirsa Irrigation Circle, India.* IWMI research report 27. IWMI, Colombo, Sri Lanka.

Bos, M.G. and Bastiaanssen, W. (2004) Using the depleted fraction to manage the groundwater table in irrigated areas. *Irrigation and Drainage Systems* 18 (3).

Bos, M.G. and Nugteren, J. (1974) *On Irrigation Efficiencies.* Publication 19. 1st edn 1974; 2nd edn 1978; 3rd rev. edn 1982; 4th edn 1990 (also published in Farsi with IRANCID, Tehran). International Institute for Land Reclamation and Improvement/ILRI, Wageningen, The Netherlands.

Bos, M.G., Abdel Dayem, S., Bastiaanssen, W. and Vidal, A. (2001) Remote sensing for water management: the drainage component. Available at: http://www.ilri.nl

Burt, C.M., Clemmens, A.J., Strelkoff, T.S., Solomon, K.H., Bliesner, R.D., Hardy, L.A., Howell, T.A. and Eisenhauer, D.E. (1997) Irrigation perfomance measures: efficiency and uniformity. *Journal of the Irrigation Drainage Engineering, ASCE* 123, 423–442.

Chambers, R. (1994) The origins and practice of participatory rural appraisal. *World Development* 22, 953–969.

Chambers, R. and Carruthers, I. (1985) Appraisal to improve canal irrigation performance: in search of cost-effective methods. In *Proceedings from the Workshop on Selected Irrigation Management Issues,* 15–19 July 1985, Digana Village, Sri Lanka. International Irrigation Management Institute research paper no. 2. International Irrigation Management Institute, Colombo, Sri Lanka.

Clyma, W. and Lowdermilk, M. (1988) *Improving the Management of Irrigated Agriculture: a Methodology for Diagnostic Analysis.* Colorado State University, Fort Collins, Colorado.

Dedrick, A.R., Bautista, E., Clyma, W., Levine, D.B., Rish, S.A. and Clemmens, A.J. (2000) Diagnostic analysis of the Maricopa-Stanfield Irrigation and Drainage District area. *Irrigation and Drainage Systems* 14, 41–67.

Gosselink, P. and Thompson, J. (1997) *Application of participatory rural appraisal methods for action research on water improvement.* Short report series on locally managed irrigation, no. 18. International Irrigation Management Institute, Colombo, Sri Lanka, 29 pp.

Kijne, C.J. (1996) *Water and Salt Balances for Irrigated Agriculture in Pakistan.* Research report 6. International Irrigation Management Institute, Colombo, Sri Lanka.

Kivumbi, D. (1999) A decision support system for improved irrigation water management. Unpublished PhD thesis, University of Southampton, UK.

Kivumbi, D., Ostrowski, M., Burton, M. and Clarke, D. (1999) Diagnostic analysis for improved irrigation performance. In ICID, *17th Congress on Irrigation and Drainage,* Granada, Spain.

Laitos, W.R., Adhikari, N.P., Banskota, B.P., Early, A.C., Gee, K.C., Johnson, D.L., Khadga, R.P., Nayman, O.B., Neupane, R.S.S., Ojha, B.P., Regmi, S.K., Shakya, C.B., Sheng, T.S., Shrestha, M.L., Yadav, S.P. and Fowler, D. (1985) *Diagnostic analysis of Sirsia irrigation system, Nepal.* WMS report no. 38, Consortium for International Development, Fort Collins, Colorado, 55 pp.

Lowdermilk, M.K., Clyma, W., Dunn, L.E., Haider, M.T., Laitos, W.R., Nelson, L.J., Sunada, D.K., Podmore, C.A. and Podmore, T.H. (1983) *Diagnostic Analysis of Irrigation Systems,* Volume 1: *Concepts and Methodology.* Colorado State University, Fort Collins, Colorado.

Merrey, D. and Baviskar, S. (eds) (1998) Gender analysis and reform of irrigation management: concepts, cases, and gaps in knowledge. *Proceedings of the Workshop on Gender and Water,* 15–19 September 1997, Habarana, Sri Lanka. IWMI, Colombo, Sri Lanka.

Merrey, D.J., Murray-Rust, D.H., Garcés-Restrepo, C., Sakthivadivel, R. and Wasanta Kumara, W.A.U. (1995) Performance improvement capacity audit. *Water Resources Development* 11, 11–23.

Molden, D. (1997) *Accounting for water use and productivity.* SWIM paper 1. IIMI, Colombo, Sri Lanka, 16 pp.

Molden, D. and Sakthivadivel, R. (1999) Water accounting to assess use and productivity of water. *International Journal of Water Resources Development* 15, 55–71.

Oad, R. and Mccornick, P.G. (1989) Methodology for assessing the performance of irrigated agriculture. *ICID Bulletin* 38, 42–53.

Perry, C.J. (1996) The IIMI water balance framework: a model for project level analysis. Research report 5. International Irrigation Management Institute, Colombo, Sri Lanka.

Podmore, C.A. and Eynon, D.G. (eds) (1984) *Diagnostic Analysis of Irrigation Systems,* Vol. 2, *Evaluation Techniques.* Water Management Synthesis Project, Colorado State University, Fort Collins, Colorado, pp. 195–238, 241–246.

Pradhan, P., Yoder, R. and Pradhan, U. (1988) Guidelines for rapid appraisal of irrigation systems: experience from Nepal. In IIMI, *Irrigation Management in Nepal.* Research papers from a National Seminar, Bharatpur, Nepal, 4–6 June 1987. IIMI pub 88–07. IIMI, Kathmandu, Nepal, pp. 24–37.

Renault, D., Hemakumara, M. and Molden, D. (2000) Importance of water consumption by perennial vegetation in irrigated areas of the humid tropics: evidence from Sri Lanka. *Agricultural Water Management* 46, 215–230.

Tuong, T.P., Cabangon, R.J. and Wopereis, M.C.S. (1996) Quantifying flow processes during land soaking of cracked rice soils. *Soil Science Society of America Journal* 60, 872–879.

Van Koppen, B. (2002) *A Gender Performance Indicator for Irrigation: Concepts, Tools and Applications.* Research report 59. IWMI, Colombo, Sri Lanka.

Vermillion, D.L. and Garcés-Restrepo, C. (1998) *Impacts of Colombia's Current Irrigation Management Transfer Program.* IIMI research report 25. IWMI, Colombo, Sri Lanka.

Vermillion, D.L., Samad, M., Pusposutardjo, S., Arif, S.S. and Rochdyanto, S. (2000) *An Assessment of the Small-Scale Irrigation Management Turnover Program in Indonesia.* IWMI research report 38. IWMI, Colombo, Sri Lanka.

Yoder, R. and Martin, E. (1985) *Identification and Utilization of Farmer Resources in Irrigation Development: a Guide for Rapid Rural Appraisal.* Irrigation Management. Network Paper 12c. ODI, London.

Further Reading

Diagnostic analysis

Bautista, E., Replogle, J.A., Clemmens, A.J., Clyma, W., Dedrick, A.R. and Rish, S.A. (2000) Water delivery performance in the Maricopa-Stanfield Irrigation and Drainage District. *Irrigation and Drainage Systems* 14, 139–166.

Dedrick, A.R., Bautista, E., Clyma, W., Levine, D.B., Rish, S.A. and Clemmens, A.J. (2000) Diagnostic analysis of the Maricopa-Stanfield Irrigation and Drainage District area. *Irrigation and Drainage Systems* 14, 41–67.

Laitos, W.R., Adhikari, N.P., Banskota, B.P., Early, A.C., Gee, K.C., Johnson, D.L., Khadga, R.P., Nayman, O.B., Neupane, R.S.S., Ojha, B.P., Regmi, S.K., Shakya, C.B., Sheng, T.S., Shrestha, M.L., Yadav, S.P. and Fowler, D. (1985) *Diagnostic Analysis of Sirsia Irrigation System, Nepal.* WMS report no. 38. Consortium for International Development, University Services Center, Colorado State University, Fort Collins, Colorado.

Podmore, C.A. and Eynon, D.G. (eds) (1984) *Diagnostic Analysis of Irrigation Systems, Volume 2: Evaluation Techniques.* Water Management Synthesis Project, Colorado State University, Fort Collins, Colorado, pp. 195–238, 241–246.

Somasekra, B.M.S., Mohammed, R.A., Imbulana, K.A.U.S., Gates, T.K., Bandaranayake, M.A.W., Nelson, L.J., Somasundera, J.W.D., Shaner, W.M., Hettige, S. and Kilkelly, M.K. (1987) *Diagnostic Analysis of Minneriya Scheme, Sri Lanka: 1986 Yala – Discipline reports.* WMS report 59. Colorado State University, Fort Collins, Colorado.

Rapid appraisal

Adams, L. (ed.) (1983) *Synthesis of Lessons Learned for Rapid Appraisal of Irrigation Systems.* Water Management Synthesis report no. 22. Utah State University, Logan, Utah.

Burt, C.M. and Styles, S.W. (1999) Modern water control and management practices in irrigation: impact on performance. FAO-IPTRID-World Bank Water Report No. 19.

Chambers, R. and Carruthers, I. (1986) *Rapid Appraisal to Improve Canal Irrigation Performance: Experience and Options.* International Irrigation Management Institute (IIMI), Digana Village, Sri Lanka.

Groenfeldt, D. (comp.) (1989) *Guidelines for Rapid Assessment of Minor Irrigation Systems in Sri Lanka.* IIMI working paper 14. IIMI, Colombo, Sri Lanka.

Harvey, J., Potten, D.H. and Schoppmann, B. (1987) Rapid rural appraisal of small irrigation schemes in Zimbabwe. *Agricultural Administration and Extension* 27, 141–155.

Khan, A.M. and Suryanata, K. (1994) *A Review of Participatory Research Techniques for Natural Resources Management.* The Ford Foundation, Jakarta, Indonesia.

Khon Kaen University (1987) *Proceedings of the 1985 International Conference on Rapid Rural Appraisal.* Khon Kaen University, Khon Kaen, Thailand.

Mabry, J.B. (ed.) (1999) *Canals and Communities: Small-scale Irrigation Systems*. University of Arizona Press, Tucson, Arizona.

Mikkelsen, B. (1995) *Methods for Development Work and Research: a Guide for Practitioners*. Sage, New Delhi.

Participatory rural appraisal

Chambers, R. (1994) Participatory rural appraisal (PRA): analysis of experience. *World Development* 22, 1253–1268.

Chambers, R. (1994) Participatory rural appraisal (PRA): challenges, potentials and paradigm. *World Development* 22, 1437–1454.

Gosselink, P. and Thompson, J. (1997) *Application of Participatory Rural Appraisal Methods for Action Research on Water Management*. Short report series on locally managed irrigation no. 18. International Irrigation Management Institute, Colombo, Sri Lanka.

Laderchi, C.R. (2001) *Participatory Methods in the Analysis of Poverty: a Critical Review*. Working paper no. 62. Queen Elizabeth House, Oxford.

Norton, A., Bird, B., Brock, K., Kakande, M. and Turk, C. (2001) *A Rough Guide to PPAs. Participatory Poverty Assessment: an Introduction to Theory and Practice*. Overseas Development Institute, London.

Selener, D., Endara, N. and Carvajal, J. (1999) *Participatory Rural Appraisal and Planning*. International Institute of Rural Reconstruction, Quito, Ecuador.

Webber, L.M. and Ison, R.L. (1995) Participatory rural appraisal design: conceptual and process issues. *Agricultural Systems* 47, 107–131.

Remote sensing

Bastiaanssen, W.G.M. (1998) *Remote Sensing in Water Resources Management: the State of the Art*. International Water Management Institute, Colombo, Sri Lanka.

Bastiaanssen, W.G.M., Thiruvengadachari, S., Sakthivadivel, R. and Molden, D.J. (1999a) Satellite remote sensing for estimating productivities of land and water. *International Journal of Water Resources Development* 15(2), 181–196.

Bastiaanssen, W.G.M., Molden, D.J., Thiruvengadachari, S., Smit, A.A.M.F.R., Mutuwatte, L. and Jayasinghe, G. (1999b) *Remote Sensing and Hydrologic Models for Performance Assessment in Sirsa Irrigation Circle, India*. Research report 27. International Water Management Institute, Colombo, Sri Lanka.

Bos, M.G. (1999) Why would we use a GIS database and remote sensing in irrigation management? *Proceedings of the 5th National ICID Symposium*, The Netherlands.

Kite, G.W. and Droogers, P. (1999) Irrigation modelling in the context of basin water resources. *Journal of International Water Resources Development* 15, 44–54.

Menenti, M., Azzali, S. and D'Urso, G. (1995) Management of irrigation schemes in arid countries. In: Vidal, A. and Sagardoy, J. (eds) *Use of Remote Sensing Techniques in Irrigation and Drainage*. Water Report 4. FAO, Rome.

Moran, M.S. (1994) Irrigation management in Arizona using satellites and airplanes. *Irrigation Science* 15, 35–44.

Neale, C.M.U., Bausch, W.C. and Heereman, D.F. (1989) Development of reflectance-based crop coefficients for corn. *Transactions of the ASAE* 32, 1891–1899.

Roerink, G.J., Bastiaanssen, W.G.M., Chambouleyron, J. and Menenti, M. (1997) Relating crop water consumption to irrigation water supply by remote sensing. *Water Resources Management* 11, 445–465.

Sakthivadivel, R., Thiruvengadachari, S., Amerasinghe, U., Bastiaanssen, W.G.M. and Molden, D.J. (1999) Performance evaluation of the Bhakra system, India, using remote sensing and GIS techniques. Research report 28. International Water Management Institute, Colombo, Sri Lanka.

Thiruvengadachari, S. and Sakthivadivel, R. (1997) Satellite Remote Sensing Techniques to Aid Irrigation System Performance Assessment: a Case Study in India. Research report 9. International Water Management Institute, Colombo, Sri Lanka.

Vidal, A. and Sagardoy, J.A. (1995) *Use of Remote Sensing Techniques in Irrigation and Drainage.* Water report 4. FAO, Rome.

Water balance and accounting

Bos, M.G. and Nugteren, J. (1982) *On Irrigation Efficiencies.* ILRI publication no. 19. ILRI, Wageningen, The Netherlands.

Burt, C.M. (2000) Irrigation water balance fundamentals. In Davids, G.G. and Anderson, S.S. (eds) *Benchmarking Irrigation System Performance Using Water Measurement and Water Balances: Proceedings from the 1999 USCID Water Management Conference,* San Luis Obispo, California, 10–13 March 1999. USCID, Denver, Colorado, pp. 1–13.

Davids, G.G. and Anderson, S.S. (eds) (2000) *Benchmarking Irrigation System Performance Using Water Measurement and Water Balances: Proceedings from the 1999 USCID Water Management Conference,* San Luis Obispo, California, 10–13 March 1999. USCID, Denver, Colorado.

Molden, D.J. (1997) *Accounting for Water Use and Productivity. SWIM (System Wide Initiative on Water Management).* Report no. 1. International Irrigation Management Institute, Colombo, Sri Lanka.

Molden, D.J., Sakthivadivel, R. and Habib, Z. (2000) *Basin Level Use and Productivity of Water: Examples from South Asia.* Research report 49. International Water Management Institute, Colombo, Sri Lanka.

Renault, D., Hemakumara, M. and Molden, D. (2000) Importance of water consumption by perennial vegetation in irrigated areas of the humid tropics: evidence from Sri Lanka. *Agricultural Water Management* 46, 215–230.

Questionnaire surveys

Beimer, P., Groves, R., Lyberg, L., Mathiowetz, N. and Sudman, S. (eds) (1991) *Measurement Errors in Surveys.* Wiley, New York.

Casley, D.J. and Kumar, K. (1987) *Project Monitoring and Evaluation in Agriculture and Rural Development Projects.* Johns Hopkins University Press, Baltimore, Maryland.

Casley, D.J. and Kumar, K. (1988) *The Collection, Analysis and Use of Monitoring and Evaluation Data.* Johns Hopkins University Press, Baltimore, Maryland.

Converse, J.M. and Presser, S. (eds) (1986) *Survey Questions: Handcrafting the Standardized Questionnaire.* Sage Publications, Beverley Hills, California.

Fink, A. (1995) *The Survey Handbook.* Sage, New Delhi.

Fink, A. and Kosecoff, J. (1985) *How to Conduct Surveys: a Step-by-Step Guide.* Sage, Beverly Hills, California.

Grosh, M. and Glewwe, P. (2000) *Designing Household Survey Questionnaires for Developing Countries: Lessons from 15 years of the Living Standards Measurement Study*, Vols 1–3. The World Bank, New York.

Podmore, C.A. and Eynon, D.G. (eds) (1983) *Diagnostic Analysis of Irrigation Systems. Vol. 2: Evaluation Techniques.* Water Management Synthesis Project, University Services Center, Colorado State University, Fort Collins, Colorado.

Shaner, W.W. and Philipp, P.F. (1982) *Farming Systems Research and Development: Guidelines for Developing Countries*, Westview Press, Boulder, Colorado.

6 Data Management for Performance Assessment

Background

Performance assessment by its nature requires data of all sorts, collected at various times and analysed by different means and presented to different audiences. With every piece of collected data, there is a potential benefit and an associated cost. A good data management programme will capture necessary information for performance assessment, maximize the utility of data collected and minimize the cost of collection and processing. In this chapter, data management implies more than merely collecting and processing data. A data management programme will also provide a means to optimize the quality of information from those of field data. The process of data management consists of two inter-related activities: (i) the design and management of the data measurement process, and (ii) the actual database management. The latter activity can be divided in a group of 'in-office' activities and a group of communication activities. The process of data management is depicted in Fig. 6.1. This chapter describes each of these phases in data management.

Data System Management

The design for cost-effective management of the data system configuration is the most critical phase in the entire data processing task. It handles key questions like 'Which data need to be measured?'

The generalized path between measured data and the presentation of performance indicators is shown in Fig. 6.2. The number of indicators that are selected for communication with end users of management information depends on the audience. A grouping of indicators is given

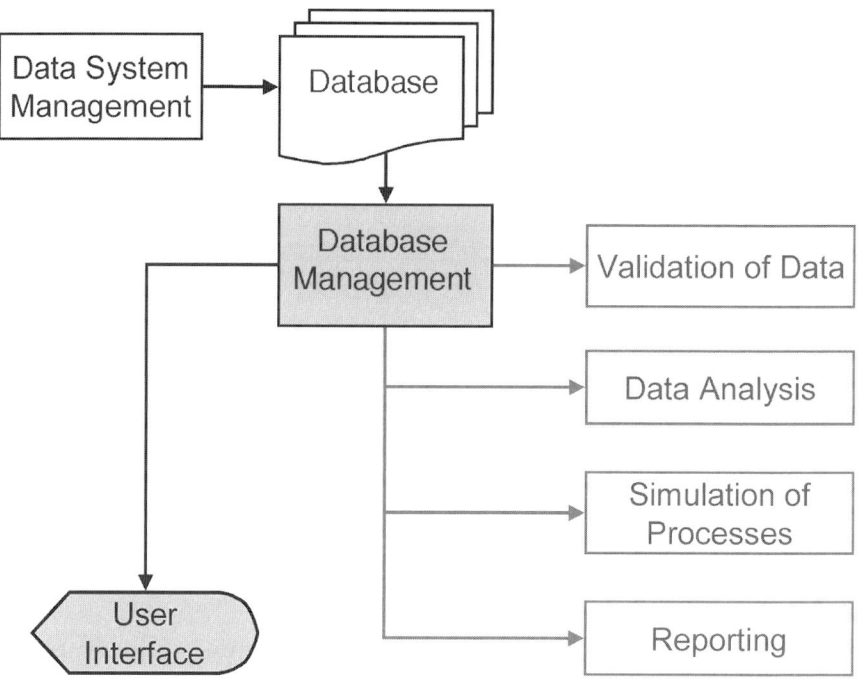

Fig. 6.1. Data processing.

in Chapter 3. The above question, 'Which data need to be measured?', also requires an answer on the density of measurement locations within the irrigated area and on the time interval between measurements.

As mentioned above, the design for cost-effective management of the data system configuration is the most critical phase in the entire data processing task. It handles key questions like:

- Which data need to be collected? This includes the density of measurement locations within the irrigated area (where?) and the time interval between measurements (how often?). The volume of data to be measured depends on the number of indicators included in the programme (see Chapter 2).
- How are data measured, e.g. should data be measured by own staff, can data be obtained or bought from other parties, can data measurement be automated in a cost-effective way?
- How are data stored? For rather straightforward assessment, an off-the-shelf spread sheet may be adequate. If the spatial distribution of indicator values needs to be reported, however, a geographic information system (GIS) is most adequate.
- How is information reported to the user of end results?

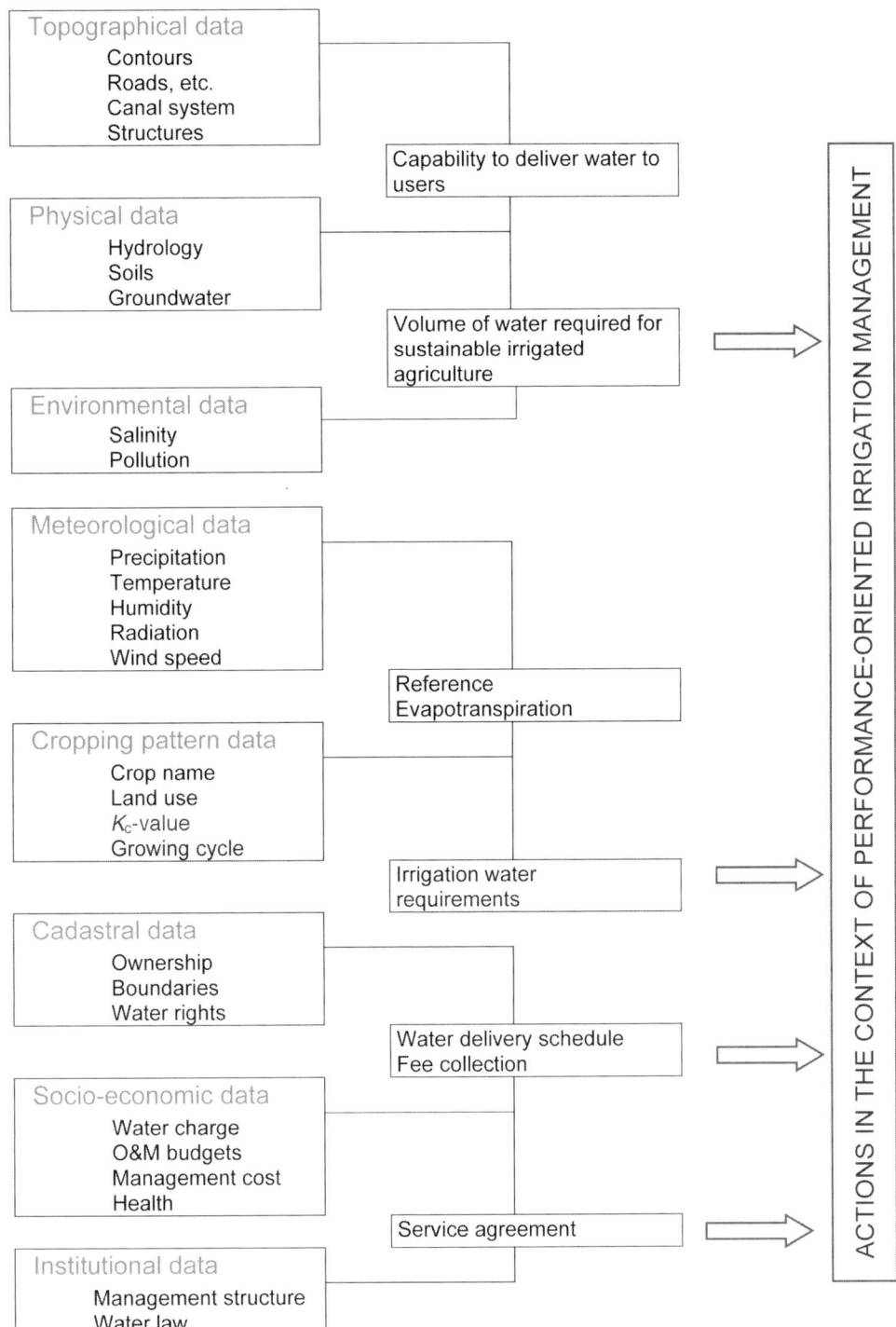

Fig. 6.2. Path from measured data to irrigation performance indicators (Bos, 2001).

Database Management

Database management can be divided into two groups of activities: (i) the in-office activities with all measured data, and (ii) the communication with end-users through 'read only' information and a user interface.

In-office activities

Before field data can be used, a thorough screening and analysis should be carried out. Thereafter, data can be used in simulation programs. These routine in-office activities lead to different styles of reports for a variety of readers (end-users). The discussion below follows the in-office activities on data management and gives general guidelines.

Validation of data

Whenever data are collected or measured, the value obtained is simply the best estimate of the true value. The true value is either slightly greater or less than the measured value. The usefulness of performance indicators is greatly enhanced if a statement of possible error accompanies the result. The error may be defined as the difference between the true value and the value that is calculated with the aid of the appropriate equations.

It is not relevant to give an absolute upper bound to the value of error. Due to chance, such bounds can be exceeded. Taking this into account, it is recommended to give a range that is expected to cover the true value of the measured quantity with a high degree of probability. This range is termed the uncertainty of measurement, and the confidence level associated with it indicates the probability that the range quoted will include the true value of the quantity being measured. A probability of 95% is commonly used as the confidence level for all errors (see 'Accuracy of Measurements and Indicators' below).

During data validation, three types of error must be considered (Fig. 6.3):

- Spurious errors as a result of human mistakes and instrument malfunctions.
- Random errors due to experimental and reading mistakes.
- Systematic errors (which may be either constant or variable).

Spurious errors are errors that invalidate a measurement. Such errors cannot be incorporated into a statistical analysis. Steps should be taken to avoid such errors and discard the results. Alternatively, corrections may be applied. Spurious errors can only be detected if time series of data are screened on irregularities and impossible values.

Random errors are errors that affect the reproducibility of measurement. It is assumed that data points deviate from the mean in accordance

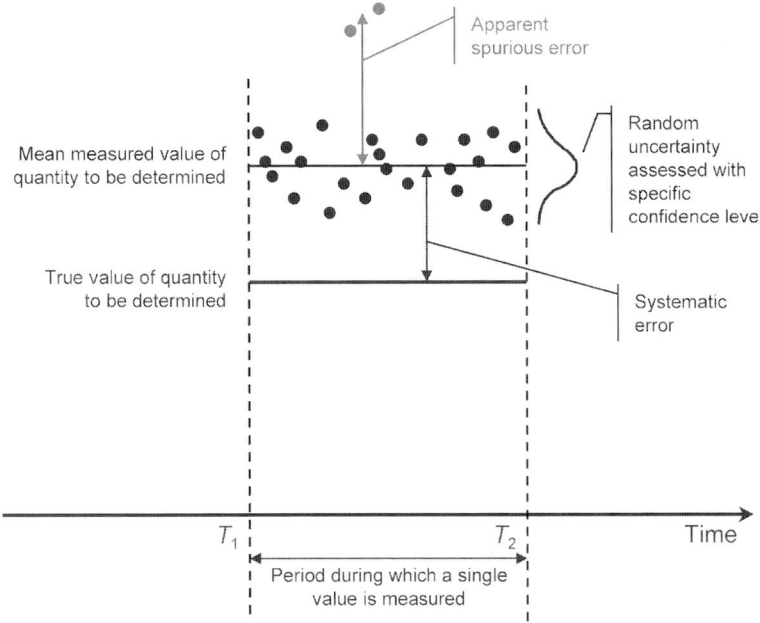

Fig. 6.3. Illustration of terms.

with the laws of chance as a result of random errors. The mean random error of a summarized measurable value over a period is expected to decrease when the number of measurements during this period increases. As a result, the integrated value over a long period of observations (more than about 15 observations) will have a mean random error that approaches zero. It is emphasized that this refers to time-dependent errors only.

Systematic errors are errors that cannot be reduced by increasing the number of measurements as long as equipment and conditions remain unchanged. Whenever there is evidence of a systematic error of a known sign, the mean error should be added to (or subtracted from) the measurement result.

Data should be validated immediately after collection in order to detect spurious and systematic errors. The cause of such errors should be corrected to avoid a gap in data during significant parts of the irrigation season. The distinction between accuracy and precision, as illustrated in Fig. 6.4, should be considered.

Data analysis

As discussed in Chapter 3, the major reason for measuring (or quantifying) the actual value of a key aspect is to see if target values of indicators are met. If the indicator value deviates too much (is outside the allowable range) of the indicator, then (corrective) adjustments need to be made. Subsequently, the impact of this corrective action on performance

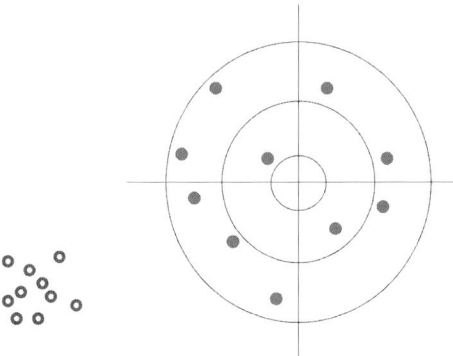

Fig. 6.4. The closed points have high accuracy and low precision; the open points have high precision and low accuracy.

needs to be monitored. Therefore, data need to be analysed and reported upon. The period between data measurement and analysis varies with the purpose of the assessment as follows:

- Operational: to adjust operation of irrigation or drainage infrastructure, data analysis is required on a real-time or near-real-time basis.
- Strategic: analysis coincides with key cycles in the irrigation or drainage process like growing season, hydrological cycles, etc. It is recommended to analyse data at least once per year (annual report on performance).
- Diagnostic: a diagnosis usually is only needed following the identification of one or more problems. Analysis is related to the nature of these problems. All data need to be analysed before a report is written.

Simulation of processes

Often the information (data) needed to enable performance-oriented management differs from the physical data that actually can be measured. The physically measured data then have to be converted into the needed data through a simulation model. For the most common processes, such models are available for use on a personal computer. For example, a flow rate (in m³/s) cannot be measured directly. However, if a weir of known dimensions is constructed, a rating equation can be derived in the shape (Clemmens *et al.*, 2001):

$$Q = K_1(h_1 - K_2)^u \tag{1}$$

in which the values of K_1, K_2 and the power u are constants for the constructed weir. If the head with respect to the weir crest, h_1, is measured, the flow rate can be calculated. If heads are recorded in digital format (Fig. 6.5), a spreadsheet can be used to transfer the heads during a selected period (e.g. day, week, month, season, etc.) into a volume of water passing the weir during this period. Parameters that can be simulated using available software are listed in Table 6.1.

Reporting

In reporting information on the performance of irrigation and drainage it is essential to review this information with respect to the average knowledge level of the reader on irrigation and drainage related processes. Thus, the same information should be reported with different terminology and different levels of detail for reader groups (water users, decision makers, system managers and researchers). Although all data can be given either in tables or in graphs, most readers prefer graphs. Researchers also may ask for data in digital format for further study.

As discussed in Chapter 3, the reporting on indicator values always is done with respect to the target value of the indicator. Depending on the purpose of the assessment (see Table 3.2), the information is presented as:

- A function of time, showing the indicator trend with respect to its target (critical) value (and the related allowable range around this target). Such a presentation in time is particularly recommended for indicators that influence crop growth (e.g. depth to groundwater, Fig. 6.6). Since 1982, the irrigation water into the area is measured and

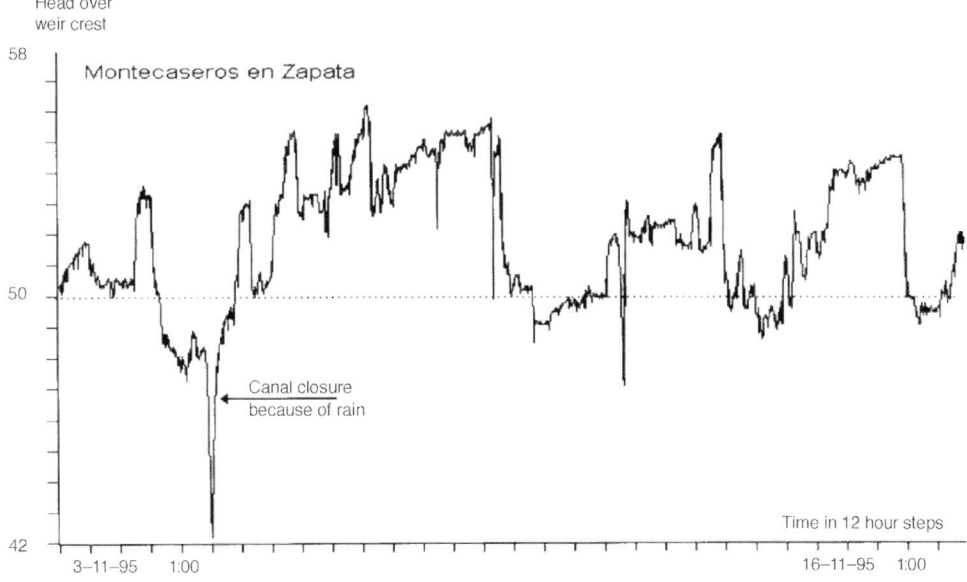

Fig. 6.5. Digital registration of head over the weir crest in an irrigation canal, Mendoza, Argentina.

Table 6.1. Overview of simulation processes for performance assessment.

Simulated parameter	Models available for simulation
Flow rate, discharge	The flow over a broad-crested weir or long-throated flume can be simulated by WinFlume (Clemmens et al., 2001). The head–discharge rating can be established with an error of 2%.
Potential evapotranspiration	The potential evapotranspiration of a cropped area can be simulated by using two methods. One method uses meteorological data and crop data to simulate ET_p. Common simulation models use the Penman–Monteith concept. Well-tested models are CROPWAT (Smith et al., 1992) and CRIWAR 2.0 (Bos et al., 1996). Because of assumptions in the theory and because of uncertainty in the used crop coefficients, the error in ET_p is about 20%. The second method estimates ET_p according to the Priestley and Taylor equation (Priestly and Taylor, 1972) using 24-h net radiation values derived from satellite data. The use of net radiation data of a particular crop under actual field conditions determined by satellites avoids the need to use generic crop coefficient data (Mekonnen and Bastiaanssen, 2000). The error in ET_p also is 20%.
Actual evapotranspiration	The actual ET from an agricultural area can be simulated from the energy balance for each pixel of a satellite image with thermal bands. Several software packages are commercially available. A well-tested program is SEBAL (Bastiaanssen et al., 1998). The error in ET_a is 20%.
Effective precipitation	Effective precipitation can be defined in various manners. The most scientifically justified method was developed by the US Department of Agriculture (1970). The method is given in CRIWAR (Bos et al., 1996). The error in the calculated effective precipitation may be as high as 20%.
Groundwater flow	Up to 100 groundwater models are available to simulate the inflow and outflow for an (irrigated) area plus the related water-level fluctuations. Some models are widely used. MODFLOW (McDonald and Harbaugh, 1988) is a popular program to simulate three-dimensional flow including the flow of chemicals in the groundwater. SIMGRO (Veldhuizen et al., 1998) was developed to simulate groundwater and surface water flow plus the water movement in the unsaturated zone. As such, it is suitable for integrated water management. MICROFEM (Hemker and Nijsten, 1997) is a semi three-dimensional program.
Soil moisture	Volumetric soil water content can be estimated empirically from satellite data, i.e. from the ratio of the latent heat flux over the net available energy fraction. The available energy then is the difference of net radiation and soil heat flux. This soil moisture value describes the average soil wetness in the root zone. If roots are absent, it describes the moisture conditions in the upper 0.05 m of the soil (Bastiaanssen et al., 1998).
Biomass production	A biomass growth routine after concepts of Asrar et al. (1985) can be used to estimate the above-ground growth of vegetation. The temporal integration of above-ground biomass growth is a good indicator of crop yield, provided that the ratio between physical harvestable yield and total biomass is known or can be established (e.g. Donald and Hamblin, 1976; Gallaghar and Biscoe, 1978).

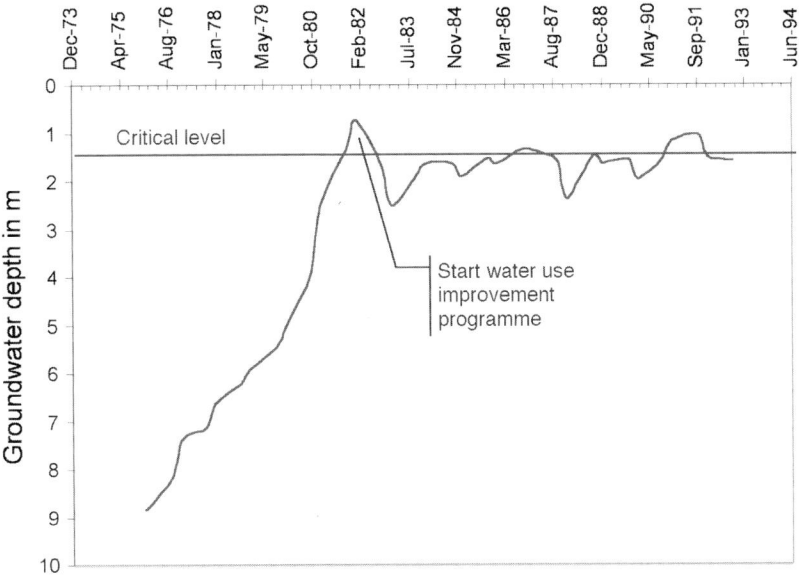

Fig. 6.6. Groundwater table data with respect to the critical level for salinity, Sirsa district, Haryana, India.

managed in such a way that the depleted fraction has an average value of about 0.6. As a result, the groundwater table remains sufficiently low to avoid salinity in the root zone.
- With an indicator value for all irrigation units (drainage areas) within the considered area. This shows the spatial distribution of the indicator. Whether the indicator value is within the allowable range or not is commonly shown with colour codes. Figure 6.7 presents information in a graphical manner. If a GIS system is used, a real scale presentation is recommended (see Figs 3.17 and 3.20).

To assess performance of irrigation and drainage (assess the use of various resources), and to decide on corrective actions in order to improve the use of these resources, the plotting of indicator values against another indicator or parameter that influences the value of the indicator is recommended. Figure 6.8 shows the impact of the depleted fraction, $ET_a/(V_c + P)$, on the fluctuation of the groundwater table. The trend line cuts the x-axis usually between 0.6 and 0.7. Thus, the water manager can influence groundwater table fluctuation by diverting another volume (V_c) of flow from the water source.

As discussed in Chapters 2 and 3, the target (intended) value of an indicator should be based on research on: 'boundary conditions' influencing the value of this indicator; critical values of the indicator that influence crop yield; and attainable (benchmark) values of the indicator that should be aimed at under similar boundary conditions. Reports on

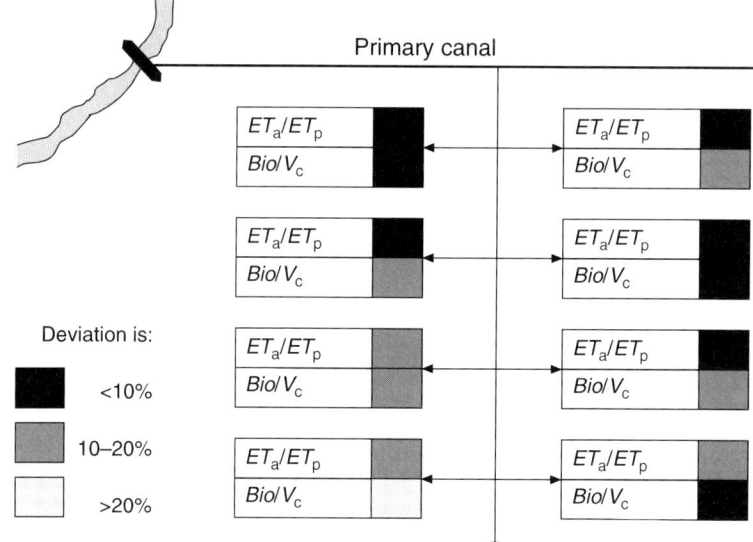

Fig. 6.7. Spatial presentation of performance indicators. ET_a/ET_p is the 'Relative evapotranspiration' and Bio/V_c is the 'Biomass production per m³ water supply' (see Chapter 3).

such research should include information on the measured mean and standard deviation of the indicator. Also, information on the allowable (operational) range should be given (Table 6.2).

User interface

For the design of a 'user interface', the first step is to define the needs of the user group. In irrigation and drainage we broadly distinguish two user groups: (i) the managers of the irrigation and drainage agency, and (ii) the customers and relations (water users, politicians, etc.) of the agency. The following concepts are recommended:

- The needs of the agency manager are directly related to the input and validation of data and with simulation processes (calculations, etc.) that produce graphs and tables. A custom-made set of screens within a commercially available spreadsheet program is recommended. Figure 6.9 (pp. 130–131) shows an example screen of such an interface.
- To facilitate communication between the agency managing irrigation or drainage and its customers (water users, public in the region, etc.) additional information is needed of better public relations quality. Besides good written information, maps are needed showing the spatial variation of indicators and parameters. To produce these maps, the

Data Management for Performance Assessment   127

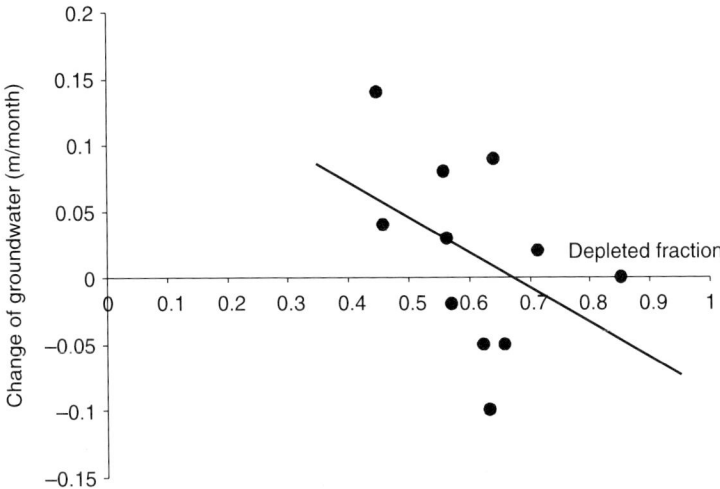

Fig. 6.8. Fluctuation of the groundwater table (in m/month) as a function of monthly averages of the depleted fraction (Nilo Coelho project, Brazil).

use of a geographic information system (GIS) is recommended. For most agencies a low-cost GIS would be adequate. An example of a GIS screen is shown in Fig. 6.10 (p. 132).

Accuracy of Measurements and Indicators

Terminology

This section gives procedures for expressing the accuracy of measured data with a randomly distributed error and the way in which these errors propagate in the calculated indicator (Bos, 1974; Clemmens, 1999). For a more detailed discussion related to irrigation, and a further reference, see Mood (1954). Examples are given on estimating accuracy where numbers are added, subtracted, multiplied or divided.

If a parameter is quantified, the obtained value is a sort of 'best measured value' of this parameter (X_1). Quantifying this parameter again through an independent measurement yields a second value (X_2) that may differ from the first measurement. If the results of many (n is more than 15) independent measurements are plotted in a histogram an envelope can be drawn around the number of measurements having a shape as shown in Fig. 6.11 (p. 133).

The average value of the measured parameter is calculated as:

$$X_{average} = \frac{\sum_{n=1}^{i} X_i}{n} \tag{2}$$

Table 6.2. Benchmark values for performance indicators for pressurized systems in irrigated fruit crops in Nilo Coelho (Brazil).

Indicator	Measured mean	Standard deviation	Operational range	Percentage of data in operational range	Acceptable range	Percentage within acceptable range
Overall consumed ratio	0.78	0.26	0.7–1.0	43	0.6–1.1	64
Depleted fraction	0.61	0.17	0.7–1.0	22	0.6–1.1	50
Crop water deficit (mm/month)	30.3	13.6	0–30	58	0–40	80
Relative evapotranspiration	0.76	0.10	0.8–1.0	35	0.7–1.0	73
Relative soil wetness	1.16	0.32	0.8–1.2	51	0.6–1.2	63
Biomass production (in kg per m^3 water supply)	2.01	1.06	>1.8	58	>1.5	58
Average				44		64

The deflection points of the envelope curve deviate ± s from the average value. The value of this 'standard deviation' can be calculated from

$$s^2 = \frac{\sum_{n=1}^{i}(X_i - X_{average})^2}{n-1} \tag{3}$$

With a normal distribution of the measured parameter, the values X_i are under the envelope curve of Fig. 6.11, while 95% of all values are within a confidence band with a width of ± 2s. A common way of expressing the 'error' of a measured parameter is by using the 'interval with a 95% confidence level'. This confidence interval (CI) is defined as:

$$CI = \pm \frac{2s}{X_{average}} \tag{4}$$

The above factor of two assumes that n is large. For $n = 6$ the factor should be 2.6; $n = 10$ requires 2.3 and $n = 15$ requires 2.1.

For example, suppose that crop cuttings (more than 15) were made to determine the yield of cotton, resulting in an average yield of 4.2 t/ha. Using a spreadsheet to calculate the standard deviation, the assessor of performance finds that $CI = 0.10$, so that the true yield falls within ± 10% of 4.2 t/ha or between 3.78 and 4.62 t/ha with 95% confidence. In other words, if cotton yield measurements under the same conditions could be repeated 100 times, 95 of the measurements would fall within ± 10% of the estimated average yield of 4.2 t/ha. Reporting 4.2 t/ha ± 10% provides much more information than just reporting the average yield. Common errors for parameters being related to irrigation and drainage are shown in Table 6.3 (pp. 133–136).

Propagation of errors

When presenting indicators, we typically take two or more parameter values, add them, multiply them and express as ratios. How do we express uncertainties in these cases?

Adding and subtracting

When adding or subtracting two values, $y = y_1 + y_2$ (or $y = y_1 - y_2$), with the confidence intervals for y_1 and y_2 being CI_1 and CI_2 respectively, an approximate estimate of the confidence interval around y expressed in terms of CI is

$$CI = \frac{\sqrt{y_{av,1}^2 CI_1^2 + y_{av,2}^2 CI_2^2}}{y} \tag{5}$$

	A	B	C	D	E	F
8		**Project Name =**	Example Project			
9		**Water Year =**	1998			
10		Total Project area (command and non-command)	100,000	Hectares; gross, including roads, all fields, water bodies		
11		Total field area in the command area	80,000	Physical area in hectares. NOT including double cropping		
12						
13		Estimated conveyance efficiency	80	Percent, %		
14		Estimated seepage for paddy rice	10	Percent, % of irrigation water delivered to fields (averaged over the irrigation season)		
15		Estimated surface losses from paddy rice to drains	10	Percent (%) of irrigation water delivered to fields		
16		Estimated field irrigation efficiency for other crops	60	Percent, %		
17						
18		Flow rate capacity of main canal(s) at diversion point(s)	70	Cubic Meters per Second (CMS)		
19		Actual Peak Flow rate into the main canal(s) at the diversion point(s)	65	Cubic Meters per Second (CMS)		
20						
21		Average ECe of the Irrigation Water	1.0	dS/m (same as mmho/cm)		
22						
23						
24						
25	This worksheet has 9 tables that require inputs FOR ONE YEAR, in addition to the cells above.					
26		Table 1 – Field Coefficients and Crop Threshold ECe				
27		Table 2 – Monthly ETo, mm				
28		Table 3 – Surface Water Entering Command Area Boundaries				
29		Table 4 – Internal Surface Irrigation Water Sources				
30		Table 5 – Hectares of Each Crop in the Command Area, by Month				
31		Table 6 – Groundwater Data				
32		Table 7 – Precipitation, effective precipitation, and deep percolation of precipitation				
33		Table 8 – Special agronomic requirements				
34		Table 9 – Crop Yields and Values				
35						

36	**Table 1 – Field Coefficients and Crop Threshold ECe**					
37			Threshold			
38			ECe			
39	Crop #	Water year month –>		Mar	Apr	May
40		**Irrigated Crop Name**	dS/m			
41	1	Paddy Rice #1	3	0.75	1.00	1.00
42	2	Paddy Rice #2	2			
43	3	Paddy Rice #3				
44	4	Crop #4	2			0.60
45	5					
46	6					
47	7					
48	8					
49	9					
50	10					
51	11					
52	12					
53	13					

Fig. 6.9. Example screen of a custom-made spreadsheet as user interface (Burt *et al.*, 2001).

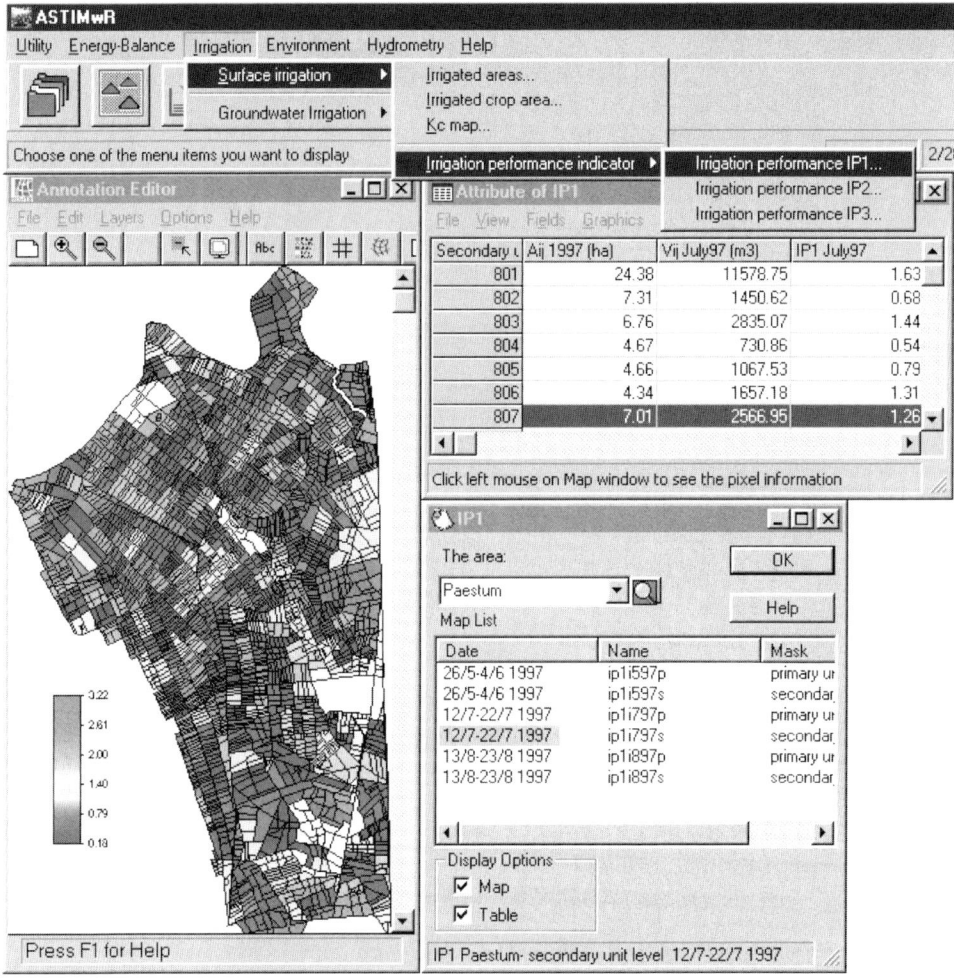

Fig. 6.10. Example of a user interface based on a GIS (ITC, 2000).

If the difference between the average values of two subtracted (independently measured) parameters is less than their standard deviation, there is a chance that one single measurement of the 'smaller' parameter exceeds the single measurement of the 'larger' parameter (Fig. 6.12). For a meaningful estimate of this difference many (more than 15) independent measurements of each parameter must be made.

Multiplication and division

For multiplication, $y = y_1 \times y_2$, an approximate estimate of the confidence interval around y expressed in terms of CI is

$$CI = \sqrt{CI_1^2 + CI_2^2 + CI_1^2 \, CI_2^2} \tag{6}$$

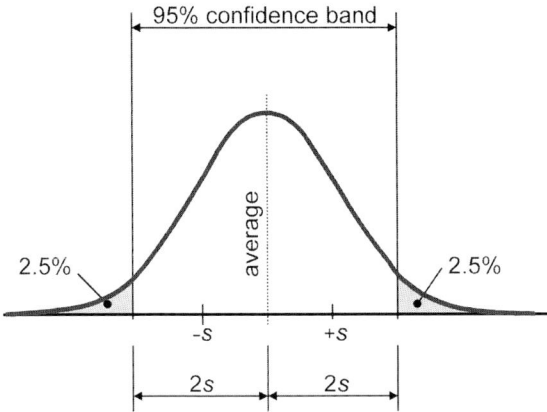

Fig. 6.11. Illustration of terminology.

Table 6.3. Definition of parameters, their methods of measurement or qualification and the estimated measurement error (95% confidence level).

Parameter	Definition	Method by which term is measured or source of data
Added mass of marketable crop	Difference between crop yield (in kg/ha) between an irrigated crop and rain-fed crop grown while the remaining conditions are the same.	Subtract crop cuttings from adjacent irrigated and rain-fed fields. Because several non-water factors are affecting yield the error exceeds 25%.
Biomass production	Total growth of vegetation (biomass added) above ground level during a selected period (day, month or season).	Subtract crop cuttings from adjacent plots at the beginning and end of the considered period. Because of spatial variation in growth the error is 15%. If biomass growth is derived from remotely sensed data the error is 20%.
Command area	Irrigable area downstream of one (considered) flow control structure.	Measured by planimetering from the most recent map of the irrigable area (5% error) or from small pixel size satellite images (5% error).
Consumption of water	Water that is actually evapotranspirated from the field and by the crop (ET_a). Consumed water enters into the atmosphere.	Point measurements can be made by lysimeters that are then extrapolated to a larger area. Remote sensing (RS) can be used to measure ET_a for a large area (for each pixel). In both cases the error is 20%.

Table 6.3. *Continued.*

Parameter	Definition	Method by which term is measured or source of data
Benchmark	The desired value of process output (or performance indicator).	The benchmark level is set by comparison with best practices of comparable processes. The set value is not subject to a statistical error.
Crop yield	Marketable yield of the cultivated crop in terms of kg/ha.	Measured by crop cuttings in the field upon harvest (error 10%).
Delivery of water	Volume of water transported (through a canal or pipe line) from a source to a customer or group of customers (water users).	If the volume of water is calculated from 15 or more individual flow measurements the error will be reduced to the systematic error in these measurements (e.g. gates 5%, weirs 2%).
Depth of delivered water	Volume of water delivered to a command area divided by the size of this area. This depth commonly has the same dimensions as precipitation and evapotranspiration, e.g. mm/day.	Is calculated as the volume of water delivered to a command area divided by the irrigated area within this command (1 mm/day = constant flow of 0.116 l/s per ha).
Design water level	Water level in a canal according to the design.	In length units and is related to a (standard) reference level. The value is not subject to a statistical error.
Discharge	Flow rate out of an area in m^3/s.	Measured by a current meter (7% error) or a flow measurement flume (long-throated flume 4% error, other structures 10% error).
Duration of water delivery	The actual duration (in time) of water delivery to an area via the structure serving this area.	Is calculated from the difference between two time readings (error 2%).
Effective precipitation	Part of precipitation that can be used to replace irrigation water.	Calculated by the US Department of Agriculture method as given in CRIWAR. The error exceeds the error for precipitation and may be about 20%.
Evapotranspiration, ET	Consumption of water by a crop and the field on which the crop is grown. This water passes into the atmosphere. ET is one process within the hydrological cycle.	Potential ET is calculated from a variety of equations. Most widely tested is Penman–Monteith (error 20%). Actual ET can be measured by lysimeter (5% error) or calculated from remote sensing data (20% error).

Table 6.3. *Continued.*

Parameter	Definition	Method by which term is measured or source of data
Fee	Money a water user has to pay to the water-delivering institution. The fee can be charged per volume delivered, per area irrigated or a combination of both.	The fee must be based on the service agreement. The value is not subject to a statistical error.
Flow rate	Volume of water passing a cross-section in a unit of time (usually second).	In m^3/s or in l/s for low flows. Measured by a current meter (7% error) or a flow measurement flume (long-throated flume 4% error, other structures 10% error).
Groundwater depth	Distance from the soil surface in the field to the groundwater level.	The groundwater depth is measured by lowering a sounder or transducer into an observation well. The random error is about 0.02 m. A systematic error of 0.05 m can occur in the ground surface elevation.
Initial irrigable area	Irrigable area at the beginning of the considered period. This period may start, for example, after completion or rehabilitation of the system.	Is determined from the design (rehabilitation) drawings of the project. The error is related to the accuracy of the map (error 1% or more).
Irrigable area	Area (in ha) with physical infrastructure that enables the delivery of irrigation water.	Is determined from the design (rehabilitation) drawings of the project. The error is related to the accuracy of the map (error 1% or more).
Irrigated area	Part of the irrigable area to which irrigation water is actually delivered during the growing season of the irrigated crop.	Is determined from collected field surveys on actual crops grown in the area (error 20%) or from satellite images with 15 or 30 m pixel size (error 5%).
Irrigation interval	The actual time in between the start of two successive water deliveries.	Is calculated from the difference between two time readings (error 1%).
Potential evapo-transpiration	$ET_{potential}$ is the evapo-transpiration by a crop that is not stressed by water shortage during its growing season.	Calculated by the equation of Penman–Monteith (CRIWAR or CROPWAT). Error is about 20%.
Regulation interval	The time interval between the start of two successive control actions for a control structure or regulator.	Is calculated from the difference between two time readings (error 2%).

Continued

Table 6.3. *Continued.*

Parameter	Definition	Method by which term is measured or source of data
Salt yield	Quantity of salts (in kg/ha) mobilized by water draining from an area. The salt yield is discharged from the area with the surface drainage water and with the groundwater.	Is calculated from the product of the flow rate and salt concentration. The surface flow is measured with a structure (error depends on structure, 2% or more) and the groundwater flow is calculated from a model (error 10% or more). Salt concentration should be measured with a modern sensor (2% error).
Service level	Amount of things provided to an organization, a project or a group of people that it needs in order to function properly and effectively.	Should be based on the (national) water law, policies or other agreements. The value is not subject to a statistical error.
Set-point	The desired value of process output (or performance indicator).	See also benchmark.
Soil moisture	The percentage (by volume) of water in the soil. If the soil is saturated it quantifies the pore space (about 40%).	Point measurements can be taken by laboratory drying of a sample (5% error) or by *in situ* electric resistance measurement (3% error). If extrapolated to a larger area, the error increases rapidly to 25%. If measured with remote sensing, the error is 20% for the considered area (including the spatial distribution).
Target value	Same as benchmark.	See also set-point.
Users participation	Participation of a water user in the (functioning of) the irrigation or drainage system.	Because 'participation' cannot be defined clearly, the error is around 40%. Thus not sufficiently accurate for performance assessment.
Volume of water	Flow rate passing a control section during a given period (day, month, season), e.g. in m^3/day, m^3/month or m^3/year.	If the volume of water is calculated from 15 or more individual flow measurements, the error will be reduced to the systematic error in these measurements (e.g. gates 5%, weirs 2%).

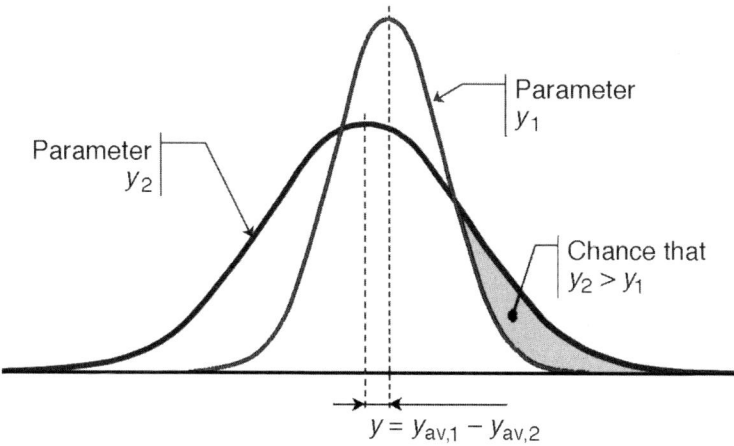

Fig. 6.12. Chance that one $y_2 > y_1$ while $y_{av,1} > y_{av,2}$.

For division, $y = y_1/y_2$, an approximate estimate of the confidence interval around y expressed in terms of CI is

$$CI = \sqrt{CI_1^2 + CI_2^2} \qquad (7)$$

Example

We want to compute the crop yield per cubic metre of water evapotranspirated by rice (the productivity). The yield of rice is based on interviews with farmers. From a statistically drawn sample, an average yield of 4000 kg/ha, $CI = 5\%$ is obtained.

An estimate of the areas under rice is 75 ha. This estimate is based on the system map and on inspection of the fields. There is uncertainty because some new houses have been built since the map was made, and it is difficult to know where some farmers have left some land fallow. So an estimate of a confidence interval of 7% is made for 'area under rice'.

Estimates of evapotranspiration for each crop in the area are based on climatic parameters following standard procedures. A value of 500 mm of ET_a for rice is obtained. There are many sources of uncertainty, including the measurement of climatic parameters, the degree of water stress during the growing season and errors associated with the means of estimating. An estimated confidence interval is set at ± 20%.

First, calculate the number of tonnes produced: 4000 kg/ha × 75 ha = 300,000 kg of rice. Substitution of the CI-values into Equation 6 gives

$$CI = \pm\sqrt{0.05^2 + 0.07^2 + 0.05^2 \times 0.07^2} = \pm 0.086$$

Then, calculate the volume of water (in m³) of evapotranspiration: 0.5 m × 75 ha = 375,500 m³. The *CI*-value of this volume is

$$CI = \pm\sqrt{0.02^2 + 0.07^2 + 0.02^2 \times 0.07^2} = \pm 0.073$$

Finally, divide the total rice yield by the cubic metres of water of evapotranspiration: productivity is 300,000/375,500 = 0.8 kg/m³. The *CI*-value is estimated through Equation 7 and gives a value of

$$CI = \pm\sqrt{0.086^2 + 0.073^2} = \pm 0.113$$

References

Asrar, G., Kanemu, E., Jackson, R.D. and Pinter, P.J. (1985) Estimation of total above-ground phytomass production using remotely sensed data. *Remote Sensing of Environment* 17, 211–220.

Bastiaanssen, W.G.M. (2000) SEBAL-based sensible and latent heat fluxes in the irrigated Gediz Basin, Turkey. *Journal of Hydrology* 229, 87–100.

Bastiaanssen, W.G.M. and Bos, M.G. (1999) Irrigation performance indicators based on remotely sensed data: a review of literature. *Irrigation and Drainage Systems* 13, 291–311.

Bastiaanssen, W.G.M., Menenti, M., Feddes, R.A. and Holtslag, A.A.M. (1998) A remote-sensing surface energy balance algorithm for land (SEBAL), part 1: formulation. *Journal of Hydrology* 212/213, 198–212.

Bos, M.G. (ed.) (1976) *Discharge Measurement Structures*. 1st edn 1976; 2nd edn 1978; 3rd rev. edn 1989. Publication 20. International Institute for Land Reclamation and Improvement/ILRI, Wageningen, The Netherlands.

Bos, M.G. (2001) Why would we use a GIS database and remote sensing in irrigation management? In: van Dijk, A. and Bos, M.G. (eds) *GIS and Remote Sensing Techniques in Land and Water Management*. Kluwer, Dordrecht, The Netherlands, pp. 1–8.

Burt, C.M. (2001) *Rapid Appraisal Process and Benchmarking*. ITRC report no. R 01–008. Funded by FAO/Thailand. Irrigation Training and Research Center, California Polytechnic State University, San Luis Obispo, California. Available at: http://www.itrc.org/reports/reportsindex.html

Clemmens, A.J. (1999) *How Accurate are Irrigation Performance Estimates? Proceedings 1999 USCID Water Management Conference: Benchmarking Irrigation System Performance Using Water Measurement and Water Balances*. San Luis Obispo, California, pp. 39–53.

Clemmens, A.J., Wahl, T.L., Bos, M.G. and Replogle, J.A. (2001) *Water Measurement with Flumes and Weirs*. Publication 58. International Institute for Land Reclamation and Improvement, Wageningen, The Netherlands. Available at: http://www.usbr.gov/wrrl/winflume

Donald, C.M. and Hamblin, J. (1976) The biological yields and harvest index of cereals as agronomic and plant breeding criteria. *Advances in Agronomy* 28, 361–405.

Gallaghar, J.N. and Biscoe, P.V. (1978) Radiation absorption, growth and yield of cereals. *Journal of Agricultural Sciences* 91, 47–60.

Hemker, C.J. and Nijsten, G.J. (1997) Ground water flow modelling using Micro-Fem, a large-capacity finite-element microcomputer package for multiple-aquifer steady-state and transient groundwater flow. Available at: http://www.microfem.com

ITC (2000) ITC Website. Available at: http://www.itc.nl/~parodi/projects.htm

McDonald, M.G. and Harbaugh, A.W. (1988) A modular three-dimensional finite-difference ground-water flow model: U.S. Geological Survey Techniques of Water-Resources Investigations, book 6, chap. A1. Available at: http://water.usgs.gov/software/modflow.html

Mekonnen, M.G. and Bastiaanssen, W.G.M. (2000) A new simple method to determine crop coefficients for water allocation planning from satellites; results from Kenya. *Irrigation and Drainage Systems* 14, 237–256.

Mood, A.M. (1954) *Introduction on the Theory of Statistics*. McGraw Hill, New York.

Priestley, C.H.B. and Taylor, R.J. (1972) On the assessment of surface flux and evapotranspiration using large-scale parameters. *Monthly Weather Review* 100, 81–92.

US Department of Agriculture (1970) *Irrigation Water Requirements*. Soil Conservation Service Technical Release 21, Washington, DC.

Veldhuizen, A.A., Poelman, A., Stuyt, L.C.P.M. and Querner, E.P. (1998) Software documentation for Simgro V3.0, Regional water management simulator. Wageningen SC-DLO, The Netherlands.

Appendix 1 Key Irrigation and Drainage System Descriptors

Code	Descriptor	Possible options (Note that this list is indicative, other options are possible)		Value
Location				
D1	Country	–		
D2	Continent	–		
D3	Scheme name	–		
D4	Latitude	–		
D5	Longitude	–		
Climate and soils				
D6	Climate	• Arid • Semi-arid • Humid	• Humid tropics • Mediterranean	
D7	Average annual rainfall (mm)	–		
D8	Average annual reference crop evapotranspiration, ET (mm)	–		
D9	Peak daily reference crop evapotranspiration, ET (mm/day)	–		
D10	Predominant soil type(s) and percentage of total area of each type	• Clay • Clay loam • Loam	• Silty clay loam • Sand	
Water source and availability				
D11	Water source	• Storage on river • Groundwater	• Conjunctive use of surface and groundwater • Run-of-the-river	
D12	Water availability	• Abundant • Sufficient	• Water scarce	

D13	Number and duration of irrigation season(s)
	Number of seasons:
	Number of months per season:
	  • Season 1:
	  • Season 2:
	  • Season 3:
		–		

Appendix 1

Code	Descriptor	Possible options (Note that this list is indicative, other options are possible)	Value
Size			
D14	Command (irrigable) area (ha)	–	
D15	Total number of water users supplied	–	
D16	Average farm size (ha)	–	
D17	Average annual irrigated area (ha)	–	
D18	Average annual cropping intensity (%)	–	
Cropping			
D19	Main crops each season with area (ha) and percentage of total command area: Crop 1: Crop 2: Crop 3: Crop 4:	–	
Institutional			
D20	Year first operational	–	
D21	Type of management	• Government agency • Private company • Joint government/local management	• Water users' association (WUA) • Federation of WUAs
D22	Agency functions	• Irrigation and drainage service • Water resources management • Reservoir management	• Flood control • Domestic water supply • Fisheries • Other
D23	Type of revenue collection	• Tax on irrigated area • Charge on crop type and area	• Charge on volume of water delivered • Charge per irrigation
D24	Land ownership	• Government	• Private
Socio-economic			
D25	(National) gross domestic product (GDP)	–	
D26	Farming system	• Cash crop • Subsistence cropping	• Mixed cash/subsistence
D27	Marketing	• Government marketing board • Private traders	• Local market • Regional/national market
D28	Pricing	• Government-controlled prices	• Local market prices • International prices
Infrastructure – Irrigation			
D29	Method of water abstraction	• Pumped diversion • Gravity diversion	• Artesian
D30	Water delivery infrastructure (length, km)	• Open channel • Pipelines	• Lined • Unlined
D31	Type and location of water control equipment	Type: • None • Fixed proportional division • Gated – manual operation • Gated – automatic local control • Gated – automatic central control	Location: • Control structure at main intake only • Control structures at primary and secondary levels • Control structures at primary, secondary and tertiary levels

Code	Descriptor	Possible options (Note that this list is indicative, other options are possible)		Value
D32	Type and location of discharge measurement facilities	Type: • Flow meter • Fixed weir or flume • Calibrated sections • Calibrated gates	Location: • None • Primary canal level • Secondary canal level • Tertiary canal level • Field level	
Infrastructure – Drainage				
D33	Area serviced by surface drains (ha)	–		
D34	Type of surface drain	• Natural	• Constructed	
D35	Length of surface drain (km)	• Natural • Constructed	• Open • Closed	
D36	Area serviced by sub-surface drainage (ha)	–		
D37	Number of groundwater level measurement sites	–		
Water allocation and distribution				
D38	Type of water distribution	• On demand • Arranged demand	• Supply-oriented	
D39	Frequency of irrigation scheduling at main canal level	• None • Daily • Weekly	• Twice monthly • Monthly • Seasonally	
D40	Predominant on-farm irrigation practice	• Surface – furrow, basin, border, flood, furrow-in-basin • Drip/trickle	• Overhead – raingun, lateral move, centre pivot • Sub-surface	

Appendix 2 Bibliography of Irrigation and Drainage Performance Indicators

The following table has been compiled from the available literature on performance assessment. It is sometimes difficult to compile such data as different authors use different terms for the same indicator. Wherever possible, the different names have been identified. Reference is made to the *Review of Selected Literature on Indicators of Irrigation Performance* by P.S. Rao (1993), who provides a valuable summary of literature on performance indicators. O&M, operation and maintenance; I&D, irrigation and drainage.

Performance indicator	Definition	Variables involved	Units	Criteria	Used by	Remarks
Water delivery and utilization						
Conveyance efficiency	$\dfrac{\text{Volume of water delivered (to tertiary unit)}}{\text{Volume of water diverted/pumped from source}}$	Discharge Duration	m³/s h	Efficiency	Bos and Nugteren (1974, 1990) Bos (1980, 1985, 1997)	Some refinement of definition between 1974 and 1997
Distribution efficiency	$\dfrac{\text{Volume of water received at field}}{\text{Volume of water delivered (to tertiary unit)}}$	Discharge Duration	m³/s h	Efficiency	Bos and Nugteren (1974, 1990) Bos (1980, 1985)	
Field application efficiency	$\dfrac{\text{Volume of water needed by crop }(ET_p - P_e)}{\text{Volume of water received at field}}$	Crop ET_p Effective rainfall, P_e Discharge Duration	mm mm m³/s h	Efficiency	Bos and Nugteren (1974, 1990) ICID (1978) Bos (1980, 1985, 1997)	Some refinement of definition between 1974 and 1997
Distribution uniformity	$\dfrac{\text{Average LQ depth irrigation water infiltrated}}{\text{Average depth infiltrated}}$	Infiltrated depth measured over an area	mm	Efficiency	Merriam and Keller (1978)	LQ – lower quartile
Irrigation system efficiency	$\dfrac{\text{Volume of water received at field}}{\text{Volume of water diverted/pumped from source}}$	Discharge duration	m³/s h	Efficiency	Bos and Nugteren (1974, 1990) ICID (1978)	
Overall project efficiency	$\dfrac{\text{Volume of water needed by crop }(ET_p - P_e)}{\text{Volume of water diverted/pumped from source}}$	Crop ET_p Effective rainfall, P_e Discharge Duration	mm mm m³/s h	Efficiency	Bos and Nugteren (1974, 1990) ICID (1978)	
Delivery performance ratio/management performance ratio	$\dfrac{\text{Actual supplied discharge}}{\text{Target discharge}}$	Actual discharge Target discharge	m³/s m³/s	Adequacy Equity Reliability	IIMI (1987) Murray-Rust and Snellen (1993) Molden and Gates (1990) Van der Velde (1990)	Used by Van der Velde to identify canal maintenance problems in Lower Chenab system
Relative water supply (RWS)	$\dfrac{\text{Total water supply}}{\text{Crop water demand}}$ Levine (1982): $\dfrac{\text{Irrigation supply + rainfall}}{\text{Seepage + Percolation} + ET_p}$	Supply discharge Duration Crop ET_p Effective rainfall, P_e	m³/s h m³ mm	Adequacy Equity	Levine (1982) Keller (1986) Weller and Payawal (1989) Bos et al. (1993, 1994) Perry (1996) Molden et al. (1998) Kloezen and Garcés-Restrepo (1998)	Widely used, and variously defined

Indicator	Formula	Data needed	Units	Category	References	Comments
Water use efficiency (WUE)	Crop water demand / Total water supply	Supply discharge, Duration, Crop ET_p, Effective rainfall, P_e	m^3/s, h, m^3, mm	Adequacy, Equity, Efficiency	Merriam et al. (1983), Merriam and Keller (1978)	Inverse of relative water supply
Relative irrigation supply	Irrigation supply / Irrigation demand $(ET_p - P_e)$	Discharge, Duration, Crop ET_p, Effective rainfall, P_e	m^3/s, h, mm, mm	Adequacy, Equity	Molden et al. (1998), Sharma et al. (1991)	Inverse of irrigation efficiency terms used by Bos and Nugteren (1974)
Reliability Index	Percentage of observations which are within ±10% of the target discharge	Actual discharge, Intended discharge	m^3/s, m^3/s		Francis (1989), Makin et al. (1990)	
Water delivery capacity (%)	Canal capacity to deliver water at system head / Peak consumptive demand	Discharge, Crop irrigation requirement	m^3/s, m^3/s	Capacity, Utility	Molden et al. (1998)	Gives an indication of the degree to which irrigation infrastructure is constraining cropping
Water distribution equity (also termed Delivery Performance Ratio, and Discharge Ratio)	Actual supply discharge / Design discharge	Actual discharge, Design discharge	m^3/s, m^3/s	Adequacy, Equity	Francis and Elawad (1989), Wolters and Bos (1990), Van der Velde (1990), Bos (1997)	Some confusion in terminology with DPR as defined below
Water delivery performance (DPR)	Actually delivered volume of water / Intended volume of delivered water	Actual discharge and duration, Intended discharge and duration	m^3/s, h, m^3/s, h	Adequacy, Equity	Lenton (1984), Molden and Gates (1990), Bos et al. (1993, 1994), Bos (1997)	Some change in terminology between 1993 and 1997 definitions
Water delivery performance error	$e^2 = 1/(n\Sigma(P_i - A_i))$	Actual water delivered (A) (at n specified locations i), Planned water delivery (P) (at n specified locations i)	m^3, m^3	Adequacy, Equity	Sharma et al. (1991)	Useful measure for assessment of a number of outlets, such as all tertiary outlets on a secondary canal
Inter-quartile ratio	Ratio of water received on best-supplied quartile of land area, to that received on worst supplied quartile	Discharge, Duration, Irrigated area	m^3/s, h, ha	Equity	Abernethy (1984), Van der Velde (1990)	
Coefficient of variation	Statistical distribution of data	Discharge, Irrigated area	m^3/s, ha	Equity	Standard, Abernethy (1984)	
Christiansen coefficient	Statistical distribution of data	Discharge, Irrigated area	m^3/s, ha	Equity	Merriam and Keller (1978), Abernethy (1984)	

Performance indicator	Definition	Variables involved	Units	Criteria	Used by	Remarks
Weekly delivery deficit	Number of weeks that water supplies are less than requirement	Water supply Water requirement	m³/s m³/s		Weller and Payawal (1989)	
Consecutive weekly delivery deficit	Number of consecutive weeks that water supplies are less than requirement	Water supply Water requirement	m³/s m³/s		Weller and Payawal (1989)	
Water availability index (WAI)	Observed water condition in paddy fields: 4.0 Water flowing paddy to paddy 3.0 Standing water in rice field 2.0 Soil is moist, in depressions 1.0 Soil is dry, surface cracks	Observed water/moisture conditions	–	Adequacy	Wijayaratne (1986)	Quoted in Murray-Rust and Snellen (1993)
Water availability index (WAI)	$\dfrac{\text{Total water supply available to scheme}}{\text{Total scheme water needs}}$	Total available water supply Total water needs	m³/year m³/year	Adequacy (input)	Ijir and Burton (1998)	Helps identify if water availability problems are due to external shortage of water or internal factors within the irrigation system Similar principle, different terminologies used by different researchers
Efficiency of infrastructure	$\dfrac{\text{Number of functioning structures}}{\text{Total number of structures}}$	No. functioning structures Total no. of structures	No. of structures	Utility Control	Mao Zhi (1989) Ijir and Burton (1998) Bos et al. (1993, 1994)	
Seepage loss ratio	$\dfrac{\text{Actual seepage rate}}{\text{Target seepage rate}}$	Seepage rate	m³/s	Efficiency	Bos et al. (1993, 1994)	
Water surface elevation	$\dfrac{\text{Actual water surface elevation at FSD}}{\text{Target water surface elevation at FSD}}$	Water surface elevation	m.a.s.l	Command/ Control	Bos et al. (1993, 1994)	
Overall reliability	$\dfrac{\text{Volume delivered}}{\text{Target volume}} \times \dfrac{\text{Actual supply duration}}{\text{Target supply duration}}$	Discharge Duration	m³/s h	Reliability	Bos et al. (1993, 1994)	
Overall consumed ratio	$\dfrac{ET_p - P_e}{\text{Volume of water diverted at intake plus other inflow}}$	Crop ET_p Effective rainfall, P_e Discharge Duration	mm mm m³/s h	Efficiency	Bos (1997)	
Conveyance ratio	$\dfrac{\text{Volume delivered to distribution system + other deliveries}}{\text{Volume diverted at intake + other inflows}}$	Discharge Duration	m³/s h	Efficiency	Bos (1997)	
Distribution ratio	$\dfrac{\text{Volume delivered to fields + other deliveries}}{\text{Volume delivered at tertiary intake}}$	Discharge Duration	m³/s h	Efficiency	Bos (1997)	

Appendix 2

Indicator	Formula	Description	Units	Category	References	Comments
Dependability of duration	$\dfrac{\text{Actual duration of water delivery}}{\text{Intended duration of water delivery}}$	Actual duration / Intended duration	h / h	Dependability	Bos (1997)	
Dependability of irrigation interval	$\dfrac{\text{Actual irrigation interval}}{\text{Intended irrigation interval}}$	Actual interval / Intended interval	days / days	Dependability	Bos (1997)	
Relative change of water level	$\dfrac{\text{Change of level}}{\text{Intended level}}$	Level change / Intended level	m / m	Command Freeboard	Bos (1997)	
Gross annual irrigation water quota	$\dfrac{\text{Total actual water delivered}}{\text{Actual irrigation area}}$	Discharge / Duration / Irrigated area	m³/s / h / ha	Adequacy	Mao Zhi (1989)	
Agricultural production						
Yield	$\dfrac{\text{Crop production}}{\text{Cropped area}}$	Crop production / Cropped area	kg / ha	Production	Standard	Influenced by many parameters, of which one is water
Relative yield	$\dfrac{\text{Actual crop yield}}{\text{Potential crop yield}}$	Estimated yield / Max. potential yield	kg/ha / kg/ha	Production	Davey and Rydzewski (1981); Abernethy (1986); Green (1989)	
Cropping intensity	$\dfrac{\text{Total area cultivated during the year}}{\text{Command area}}$	Total cropped area / Command area	ha / ha	Production	Standard	Fundamental indicator of scheme performance
Area utilization	$\dfrac{\text{Harvested area}}{\text{Theoretically serviceable area}}$	Harvested area / Service area (command area)	ha / ha	Production Efficiency (of land use)	Garces (1983)	
Specific yield/water use efficiency (kg/m³)	$\dfrac{\text{Crop production}}{\text{Total volume of water supplied in season}}$	Crop yield / Water supplied	kg / m³	Efficiency Productivity	ICID (1978); Garcés (1983); Weller and Payawal (1989); Mao Zhi (1989)	Easier to use with mono-culture
Relative productivity of water	$\dfrac{\text{Potential crop production}}{\text{Total water supplied}}$	Potential crop yield / Water supplied	kg/ha / m³/ha	Efficiency Productivity	Davey and Rydzewski (1981); Abernethy (1986); Green (1989); Mao Zhi (1989)	Similar principle, different terminologies used by different researchers
Relative crop planting dates	Variation (in days) from optimum crop planting dates	Crop planting date	date		Weller and Payawal (1989); Tiffen (1990); Ijir and Burton (1998)	Similar principle, different terminologies used by different researchers

Performance indicator	Definition	Variables involved	Units	Criteria	Used by	Remarks
Annual yield	$\dfrac{\text{Annual crop production}}{\text{Command area}}$	Annual crop production Command area	kg ha	Production	General use Abernethy (1990)	Clearer with monoculture
Output per cropped area ($/ha)	$\dfrac{\text{Value of production}}{\text{Irrigated cropped area}}$	Crop production Crop market price Irrigated crop area	kg/ha $/kg ha	Production	Molden et al. (1998) Kloezen and Garcés-Restrepo (1998)	
Output per unit command ($/ha)	$\dfrac{\text{Value of production}}{\text{Command area}}$	Crop production Crop market price Command area	kg/ha $/kg ha	Production	Molden et al. (1998) Kloezen and Garcés-Restrepo (1998)	
Output per unit irrigation supply ($/m³) (water productivity)	$\dfrac{\text{Value of production}}{\text{Diverted irrigation supply}}$	Crop yield Crop market price Crop area Supply discharge	kg/ha $/kg ha m³/s	Production	Molden et al. (1998) Kloezen and Garcés-Restrepo (1998)	
Output per unit water consumed ($/m³) (water productivity)	$\dfrac{\text{Value of production}}{\text{Volume of water consumed by ET}}$	Crop yield Crop market price Crop area Actual crop ET	kg/ha $/kg ha mm	Production	Molden et al. (1998) Kloezen and Garcés-Restrepo (1998)	
Irrigated area performance	$\dfrac{\text{Actual area}}{\text{Target area}}$	Crop area	ha	Utility	Mao Zhi (1989) Bos et al. (1993, 1994)	
Cropping intensity performance	$\dfrac{\text{Actual cropping intensity}}{\text{Target cropping intensity}}$	Crop areas	ha	Utility	Mao Zhi (1989) Bos et al. (1993, 1994)	
Production performance	$\dfrac{\text{Total production}}{\text{Target production}}$	Crop types Crop yields Crop areas	– kg/ha ha	Production	Bos et al. (1993, 1994)	
Yield performance	$\dfrac{\text{Actual yield}}{\text{Target yield}}$	Crop yield	kg/ha	Production	Bos et al. (1993, 1994)	
Water productivity performance	$\dfrac{\text{Actual water productivity}}{\text{Target water productivity}}$	Crop type Crop area Crop yield Actual water supply Target water supply	– ha kg/ha m³ m³	Productivity	Bos et al. (1993, 1994)	

Appendix 2

Agricultural economic and financial

Indicator	Formula	Units	Category	Standard	Comments
Profitability	Farm income minus expenditure		Profitability		
Resource utilization	Value of production / Cost of production	kg/ha $/kg $/kg	Efficiency	Abernethy (1990)	Influenced by many factors
Fee collection index (also fee collection performance)	Irrigation fees collected / Irrigation fees due	$ $ No. $ $	Efficiency Sustainability	Garcés (1983) Abernethy (1990) Bos et al. (1993, 1994) Bos (1997) Ijir and Burton (1998)	
Gross return on investment (%)	Standardized gross value of production / Cost of irrigation infrastructure	kg/ha $/kg ha $	Productivity Efficiency	Molden et al. (1998)	
Financial self-sufficiency	Revenue from irrigation / Total O&M expenditure	$ $	Financial viability	Molden et al. (1998) Kloezen et al. (1997) Bos (1997) Ijir and Burton (1998)	Similar principle, different terminologies used by different researchers
Total financial viability	Actual O&M allocation / Required O&M allocation	$ $	Financial viability	Garcés (1983) Mao Zhi (1989) Bos et al. (1993, 1994) Ijir and Burton (1998)	Similar principle, different terminologies used by different researchers
Income from water charges per unit area ($/ha)	Revenue from I&D charges / Command area	$ ha	Financial viability Sustainability	Mao Zhi (1989) Kloezen et al. (1997)	Varies for different systems, but a useful broad indicator nevertheless
Area-based profitability	Incremental benefit per unit area / Total irrigation expenses	$ ha $	Profitability	Mao Zhi (1989) Bos et al. (1993)	
Water-based profitability	Incremental benefit per unit water / Total irrigation expenses	$ m³ $	Profitability	Mao Zhi (1989) Bos et al. (1993, 1994)	
O&M fraction	Cost of operation + maintenance / Total agency budget	$ $	Operational viability	Bos (1997)	

Performance indicator	Definition	Variables involved	Units	Criteria	Used by	Remarks
Yield vs. water cost ratio	$\dfrac{\text{Added value of crop}}{\text{Cost of applied irrigation water}}$	Value irrigated crop Value rain-fed crop Cost of applied water	$ $ $	Profitability	Bos (1997)	
Yield vs. water supply ratio	$\dfrac{\text{Added mass of marketable crop}}{\text{Mass of irrigation water delivered}}$	Mass of irrigated crop Mass of rain-fed crop Mass of irrigation water	kg kg kg	Productivity	Bos (1997)	
Irrigation benefit per unit area	Benefit from irrigated crops − benefit from crops without irrigation − costs of irrigation	Irrigated crop yield Non-irrigated crop yield Crop market price Costs of irrigation	kg/ha kg/ha $/kg $/ha	Productivity	Mao Zhi (1989)	
Socio-economic						
Quality of life	Can vary widely	Public health Standard of living Employment levels, etc.		Quality	Chambers (1988) Abernethy (1990)	Very difficult to measure and set standards
Farmers' satisfaction	The degree of satisfaction perceived by the farmers with the level of service provision	Farmer perception (obtained through questionnaire survey)	–	Satisfaction	Garces (1983)	Should be more widely used
Irrigation employment generation	$\dfrac{\text{Annual person days per ha labour in scheme}}{\text{Annual number official working days}}$	Total person days labour Total area Number of annual working days	No. ha No.	Employment	Chambers (1988) Bos et al. (1993, 1994)	
Irrigation wage generation	$\dfrac{\text{Average rural income}}{\text{Annual national (regional) average income}}$	Average rural income Average national income	$/year $/year	Income generation	Bos et al. (1993, 1994)	
Relative poverty	$\dfrac{\text{Percent population above poverty line (scheme)}}{\text{Percent population above poverty line (national)}}$	Poverty line income Numbers earning and income levels (scheme and nationally)	$/year No. $/year	Livelihood	Bos et al. (1993, 1994)	
Technical knowledge of staff	$\dfrac{\text{Knowledge required to fulfil job}}{\text{Actual knowledge of staff}}$	–	–		Bos et al. (1993, 1994) Bos (1997) Ijir and Burton (1998)	Similar principle, different terminologies used by different researchers

Appendix 2

Indicator	Formula	Units	Category	Reference	Remarks
Users' stake in irrigation system	$\dfrac{\text{Active water users' organizations}}{\text{Total number of water users' associations}}$	–		Bos et al. (1993, 1994); Bos (1997)	
Response capacity	Measure of the ability of the O&M agency staff to address day-to-day O&M issues	O&M agency staff capabilities, attitudes and responsiveness	Efficiency Efficacy	Garcés (1983)	
Manpower numbers ratio	$\dfrac{\text{Total O\&M staff numbers}}{\text{Total irrigable area}}$	No. ha	Efficiency (staffing)	Ijir and Burton (1998)	
Scheme development ratio	$\dfrac{\text{Total scheme area actually developed for irrigation}}{\text{Total potential development area}}$	$\dfrac{\text{Actual area irrigable}}{\text{Potential area irrigable}}$ ha ha	Utility	Ijir and Burton (1998)	
Environment					
Sustainability of irrigated area	$\dfrac{\text{Current irrigable area}}{\text{Initial irrigated area}}$	Irrigable area ha	Utility	Bos et al. (1993, 1994); Bos (1997); Ijir and Burton (1998)	Similar principle, different terminologies used by different researchers
Irrigation and drainage water quality	Water quality measured against water quality standards	EC, BOD, SAR, etc. –	Quality	Standard	
Relative groundwater depth	$\dfrac{\text{Actual groundwater depth}}{\text{Critical groundwater depth}}$	Actual groundwater depth Critical groundwater depth m m	Sustainability	Bos (1997)	
Relative EC ratio	$\dfrac{\text{Actual EC value}}{\text{Critical EC value}}$	Actual EC value Critical EC value	Sustainability	Bos (1997)	
Waterlogging index	$\dfrac{\text{Area affected by waterlogging}}{\text{Total command area}}$	Total waterlogged area Command area ha ha	Productivity Sustainability	Garcés (1983)	

ICID Performance Assessment Guidelines

List of References Associated with Appendix 2

Abernethy, C.L. (1984) *Methodologies for Studies of Irrigation Water Management*. Report OD/TN 9, October. Hydraulics Research, Wallingford, UK.

Abernethy, C.L. (1986) *Performance Measurement in Canal Water Management: a Discussion*. ODI/IIMI Irrigation Management Network Paper 86/2d. Overseas Development Institute, London.

Abernethy, C.L. (1990) Indicators and criteria of the performance of irrigation systems. Paper presented at the *FAO Regional Workshop on Improved Irrigation System Performance for Sustainable Agriculture*, Bangkok, Thailand, 22–26 October.

Bos, M.G. (1980) Irrigation efficiencies at crop production level. *ICID Bulletin* 29, July.

Bos, M.G. (1985) Summary of ICID definitions on irrigation efficiency. *ICID Bulletin* 34, January, pp. 28–31.

Bos, M.G. (1997) Performance assessment indicators for irrigation and drainage. *Irrigation and Drainage Systems* 11, 119–137.

Bos, M.G. and Nugteren, J. (1974) *On Irrigation Efficiencies*, ILRI publication no. 19. International Institute for Land Reclamation and Improvement (ILRI), Wageningen, The Netherlands.

Bos, M.G. and Nugteren, J. (1990) *On Irrigation Efficiencies*, 2nd edn. ILRI publication no. 19. International Institute for Land Reclamation and Improvement (ILRI), Wageningen, The Netherlands.

Bos, M.G., Murray-Rust, D.H., Merrey, D.J., Johnson, H.G. and Snellen, W.B. (1993) Methodologies for assessing performance or irrigation and drainage management. Paper presented at the *Workshop of the Working Group on Irrigation and Drainage Performance, ICID 15th International Congress*, The Hague, The Netherlands.

Bos, M.G., Murray-Rust, D.H., Merrey, D.J., Johnson, H.G. and Snellen, W.B. (1994) Methodologies for assessing performance of irrigation and drainage management. *Irrigation and Drainage Systems* 7, 231–262.

Chambers, R. (1988) *Managing Canal Irrigation: Practical Analysis from South Asia*. Cambridge University Press, Cambridge.

Davey, C.J.N. and Rydzewski, J.R. (1981) Evaluation of water management on irrigation projects. *12th ICID Congress*, Special Session, Grenoble.

Francis, M.R.H. (1989) *Research for Rehabilitation: Study of Reliability of Water Supply to Minor Canals*. Interim report no. EX1981 (restricted circulation). Hydraulics Research, Wallingford, UK.

Francis, M.R.H. and Elawad, O. (1989) Diagnostic investigations and rehabilitation of canals in the Gezira irrigation scheme, Sudan. *Asian Regional Symposium on the Modernisation and Rehabilitation of Irrigation and Drainage Schemes*. Hydraulics Research, Wallingford, UK.

Garcés, C. (1983) A methodology to evaluate the performance of irrigation systems: application to Philippine national systems. Unpublished PhD thesis, Cornell University, Ithaca, New York.

Green, A.P.E. (1989) *A Productivity Indicator for Paddy Rice*. Report ODU 45 (Draft). Hydraulics Research, Wallingford, UK.

ICID (1978) Standards for the calculation of irrigation efficiencies. *ICID Bulletin* 27.

IIMI (1987) *Study of Irrigation Management – Indonesia*. Final report. International Irrigation Management Institute, Colombo, Sri Lanka.

Ijir, T.A. and Burton, M.A. (1998) Performance assessment of the Wurno Irrigation Scheme, Nigeria. *ICID Journal* 47, 31–46.

Keller, J. (1986) Irrigation system management. In: Node, K.C. and Sampath, R.K. (eds) *Irrigation Management in Developing Countries: Current Issues and Approaches*. Studies in Water Policy and Management No. 8. Westview Press, Boulder, Colorado, pp. 329–352.

Kloezen, W.H. and Garcés-Restrepo, C. (1998) *Assessing Irrigation Performance with Comparative Indicators: the Case of the Alto Rio Lerma Irrigation District, Mexico*. Research report 22. International Water Management Institute, Colombo, Sri Lanka.

Kloezen, W.H., Garcés-Restrepo, C. and Johnson, S.H. III (1997) *Impact Assessment of Irrigation Management Transfer in the Alto Rio Lerma Irrigation District, Mexico*. Research report 15. International Water Management Institute, Colombo, Sri Lanka.

Lenton, R.L. (1984) A note on monitoring productivity and equity in irrigation systems. In: Pant, N. (ed.) *Productivity and Equity in Irrigation Systems*. Ashish Publishing House, New Delhi.

Levine, G. (1982) *Relative Water Supply: An Explanatory Variable for Irrigation Systems*. Technical report no. 6. Cornell University, Ithaca, New York.

Makin, I.W., Goldsmith, H. and Skutsch, J.D. (1990) *Ongoing Performance Assessment – a Case Study of Kraseio Project, Thailand*. Report OD/P 96. Hydraulics Research, Wallingford, UK.

Mao Zhi (1989) *Identification of Causes of Poor Performance of a Typical Large-sized Irrigation Scheme in South China*. ODI/IIMI Irrigation management network paper 89/1b. Overseas Development Institute, London, June.

Merriam, J.L. and Keller, J. (1978) *Farm Irrigation System Evaluation: a Guide to Management*. Utah State University, Logan, Utah.

Merriam, J.L., Shearer, M.N. and Burt, C.M. (1983) Evaluating irrigation systems and practices. In: Jensen, M.E. (ed.) *Design and Operation of Farm Irrigation Systems*. ASAE monograph no. 3. American Society of Agricultural Engineers, St Joseph, Michigan.

Molden, D.J. and Gates, T.K. (1990) Performance measures for evaluation of irrigation water delivery systems. *Journal of Irrigation and Drainage Engineering, ASCE* 116, 804–823.

Molden, D.J., Sakthivadivel, R., Perry, C.J., de Fraiture, C. and Kloezen, W. (1998) *Indicators for Comparing Performance of Irrigated Agricultural Systems*. Research report 20. International Water Management Institute, Colombo, Sri Lanka.

Murray-Rust, D.H. and Snellen, W.B. (1993) *Irrigation System Performance Assessment and Diagnosis*. Joint IIMI/ILRI/IHEE publication. International Irrigation Management Institute, Colombo, Sri Lanka.

Perry, C.J. (1996) Quantification and measurement of a minimum set of indicators of the performance of irrigation systems. Mimeo, International Irrigation Management Institute, Colombo, Sri Lanka.

Rao, P.S. (1993) *Review of Selected Literature on Indicators of Irrigation Performance*. IIMI research paper. International Irrigation Management Institute, Colombo, Sri Lanka.

Sharma, D.N., Ramchand Oad and Sampath, R.K. (1991) Performance measures for irrigation and water delivery systems. *ICID Bulletin* 40(1), 21–37.

Tiffen, M. (1990) Variability in water supply, incomes and fees: illustrations of viscious circles from Sudan and Zimbabwe. *ODI/IIMI Irrigation Management Network Paper* 90/1b, April.

Van der Velde, E.J. (1990) Performance assessment in a large irrigation system in Pakistan: opportunities for improvement at the distributary level. Paper presented at the *FAO Regional Workshop on Improved Irrigation System Performance for Sustainable Agriculture*, Bangkok, Thailand, 22–26 October.

Weller, J.A. and Payawal, E.B. (1989) *Performance Assessment of the Porac Irrigation Systems*. Report OD-P 74. Hydraulics Research, Wallingford, UK.

Wijayaratne, C.M. (1986) Assessing irrigation system performance: a methodological study with application to the Gal Oya System, Sri Lanka. Unpublished PhD thesis, Cornell University, Ithaca, New York.

Wolters, W. and Bos, M.G. (1990) Irrigation performance assessment and irrigation efficiency. *1989 Annual Report*. International Institute for Land Reclamation and Improvement, Wageningen, The Netherlands.

Index

accountability assessment 10
accounting, water 106–108, 109(tab)
accuracy
    errors 120–121, 129, 132–138
    vs. precision 122(fig)
    terminology 127, 129
actual values 29(tab), 30
agreements, service *see* services, provision of
Aix-en-Provence, France 79(tab)
analysis, data 121–122
applications of performance assessment 3–4
appraisal
    participatory rural 99
    rapid 98–99, 100–101(box)
Argentina 34(fig), 42(fig), 53(fig), 123(fig)
Australia 68–69(tab), 79(tab)

balance, water *see* water balance
benchmarking 11, 29(tab), 128(tab)
Bhakra irrigation system, India 105(fig)
biomass yield over water supply ratio 56–57
boundaries, assessment 13–14
Brazil *see* Nilo Coelho project, Brazil
Burkina Faso 106

canals
    schematic of system 39(fig)
    water level and head–discharge 43–45
cause and effect relationships 102, 103(fig)
Chishtian, Pakistan 107–108
Colombia 96(fig)
competition for water 1–2, 106
confidence intervals 129, 132–138

consumption vs. use 46
control systems 72–73
conveyance *see* outflow over inflow ratios
costs, relative 54
criteria, assessment
    according to type of person 17(tab)
    linkage with performance indicators 17, 19(tab)
    selection 11, 15, 77, 79
critical values 28(fig), 29(tab), 30
crop water deficits 54–55
cropped area ratios 47–48
crops
    price ratios 54
    yields 49, 50, 51(fig)

data
    analysis 121–122
    collection 19–21, 22–23(tab)
    data system management 117–119
    for diagnostic analysis 94–95, 96(fig)
    linkage to performance indicators 20(tab), 23(tab)
    processing and analysis 80
        example 81(tab), 82(fig)
    from remote sensing 33(tab), 54–57, 103–104, 105(fig)
    in reports 123, 125–126
    simulation models 122–123, 124(tab)
    user interfaces 126–127
    validation 120–121
    see also accuracy
delivery performance ratios 40–41, 42(fig), 82(fig)

dependability: of irrigation intervals 41–43
depleted fraction ratios 36–38, 38(fig)
descriptors, key 12–13(tab)
diagnostic analysis 10
    characteristics 88(box)
    concepts and principles 90–92
    data handling 122
    methodologies 97–103
    in operational context 87–89
    reasons to perform 89, 90(tab)
    six-step approach 92–97
    specific techniques 103–110
    who does it 89
diagnostic trees 99, 102, 103(fig)
dimensionless performance indicators 28(fig)
discharge capacity ratios 44
DPR *see* delivery performance ratios
drainage ratios 38, 39(tab)

economics-related indicators 48–54
efficiency *see* outflow over inflow ratios
Egypt 44(fig)
environment-related indicators 32–33(tab), 45–48
errors *see* accuracy
evapotranspiration 54–55, 56(fig)
external vs. internal assessment 11

fees: collection ratios 52
field application ratios 35–36, 46
financial viability 50–52, 53(fig)
France 68–69(tab), 79(tab)

Gediz Basin, Turkey 50(fig)
gender performance indicators 104, 106
generic frameworks
    structure 6–7, 8(fig)
geographic information systems (GIS) 127, 132(fig)
Goulburn-Murray Irrigation District, Australia 68–69(tab), 79(tab)
groundwater 125(fig), 127(fig)
    depth 45–46

head–discharge 43–45
hierarchies: of objectives 16
hypotheses, working 94

implementers of assessments 9–10
India 68–69(tab), 78(tab), 105(fig), 125(fig)

indicators, performance
    benchmark values 128(tab)
    desirable properties 27(tab)
    in diagnostic analysis 92
    gender performance 104, 106
    linkage with criteria 17, 19(tab)
    linkage with data collection 20(tab), 23(tab)
    major functions 58–59(tab)
    measurements 119(fig), 133–136(tab)
    parameter critical value 28(fig), 29(tab)
    for reports on water users' associations 83–84(tab)
    selection 77, 79
    spatial presentation 126(fig)
    terminology 29(tab), 30
    types 20(tab), 30–31, 58–59(tab)
        dimensionless indicators 28(fig)
        economics-related 33(tab), 48–54
        environment-related 32–33(tab), 45–48
        from remote sensing 33(tab), 54–57
        water-related 31, 32(tab), 33–45, 82(fig)
Indonesia 78(tab), 110(fig)
infrastructure
    components 70(fig)
    effectivity 44
    importance 64–65
intended values 29(tab), 30
interfaces, user 126–127, 130–131(fig), 132(fig)
internal vs. external assessment 11
intervals, irrigation 41–43
intervention assessment 10

key parameter service levels 29(tab)
Kirindi Oya system, Sri Lanka 108, 109(tab)

land, productivity of 49–50
levels, canal water 43–45
Los Sauces, Mendoza, Argentina 42(fig)

management 73
    data systems 117–119
    databases 120–127
    irrigation management cycle 75(fig)
    operational management cycle 74(fig)
mass balance equations 107
Mendoza, Argentina 42(fig), 123(fig)
models, simulation 34, 47, 122–123, 124(tab)
Mogambo Irrigation Scheme, Somalia 16(tab)
MO&M (management, operation and maintenance) costs 50–52
Morocco 68–69(tab)

Nepal 78(tab)
Nilo Coelho project, Brazil 38(fig), 40(fig), 46(fig), 49(fig), 51(fig), 52, 55–57, 56(fig), 95(fig), 127(fig), 128(tab)
nitrates 47(tab)

objectives 2–3
    linkage with criteria, indicators and targets 15–16
    objective-setting 7, 9(tab), 16–17
    for state farm and settlement scheme 18(tab)
O&M (operation and maintenance)
    budgets and self-sufficiency 96(fig)
    fractions 51–52, 53(tab)
    staff salaries 102
operational management cycle: strategic planning 74(fig)
operational performance 3, 10
    action after assessment 85
    assessment 75–80, 87–89, 122
organic matter 47(tab)
outflow over inflow ratios 38–39, 40(fig), 43
outputs, report 21, 24
overall consumed ratios 34–35
oxygen, demand for 47(tab)

Pakistan 107–108
parameters see indicators, performance
participatory rural appraisal 99
performance
    indicators see indicators, performance
    operational see operational performance
    strategic see strategic performance
phosphorus 47(tab)
pipes: performance quantification 44
pollution, water 46–47
precision vs. accuracy 122(fig)
price ratios 54
productivity, water and soil 48–50, 56, 106

quality, water 67(tab)
questionnaire surveys 108–110
questions: diagnostic analysis 93–94

ranking: of objectives 16–17, 18(tab)
rapid appraisal 98–99, 100–101(box)
ratios as performance indicators
    biomass yield over water supply 56–57
    cropped area 47–48
    delivery performance 40–41, 42(fig), 82(fig)
    depleted fraction 36–38, 38(fig)
    discharge capacity 44
    drainage 38, 39(tab)
    fee collection 52
    field application 35–36, 46
    MO&M funding 51
    outflow over inflow 38–39, 40(fig), 43
    overall consumed 34–35
    of prices 54
    water level 44
relative evapotranspiration 55, 56(fig)
relative soil wetness 56, 57(fig)
remote sensing 33(tab), 54–57, 103–104, 105(fig)
reports 83–84, 123, 125–126
river basins 38, 39(tab), 50(fig)

salinity, soil 47(tab)
satellites see remote sensing
schedules, irrigation provision 65–66
services, provision of 63–67, 68–69(tab), 70(fig), 72–73
Shereishra Pilot Area, Egypt 44(fig)
simulation, process 34, 47, 122–123, 124(tab)
Sirsa district, Haryana, India 125(fig)
Societe du Canal de Provence, France 68–69(tab)
soil
    relative wetness 56, 57(fig)
    salinity 47(tab)
Somalia 16(tab)
specifications: service provision 65–67, 68–69(tab)
Sri Lanka 108, 109(tab)
stakeholders 7, 9, 76, 91
    reporting to 83–84
standards: water quality 67(tab)
strategic performance 3
    action after assessment 85(fig), 86
    assessment 74–75, 76, 77, 80, 122
    and diagnostic analysis 88(fig), 89
sustainability 10, 47–48
systems approach 13–14, 91
    nested systems 18 (fig)
systems, irrigation: components 70, 71(fig)

target values 29(tab)
total values 29(tab), 30
Trigga Scheme, Morocco 68–69(tab)
Tunayan, Argentina 34(fig), 53(fig)
Turkey 50(fig)

use vs. consumption 46
user interfaces 126–127, 130–131(fig), 132(fig)

validation, data 120–121

Warabandi, India 68–69(tab), 78(tab)
water balance
    and accounting 106–108, 109(tab)
        schematic representation of flows 37(fig)
water productivity 48–49, 50(fig), 56
water quality standards 67(tab)
water-related indicators 31, 32(tab), 33–45

water users' associations
    assessment and reports 83–84(tab)
    rapid appraisal checklist 100–101(box)
    water allocation 81(tab)
weighting: of objectives 16–17, 18(tab)
wetness, relative soil 56, 57(fig)
working hypotheses 94

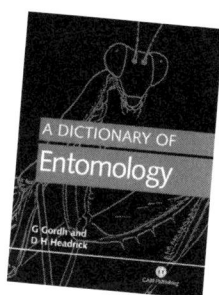